Deane's Doctrine of Naval Architecture, 1670

Deane's Doctrine of Naval Architecture, 1670

Edited and introduced by Brian Lavery

CONWAY MARITIME PRESS

Introduction, transcription and notes © Brian Lavery 1981
© Conway Maritime Press Ltd 1981

First published in 1981 by
Conway Maritime Press Ltd
2 Nelson Road, Greenwich, London SE10 9JB

ISBN 0 85177 180 7

Designed by Jon Blackmore and Ray Fishwick
Jacket designed by Dave Mills
Typesetting by The Manor Studio, London
Printed and bound in Great Britain by Butler & Tanner, Frome

CONTENTS

Introduction	7
Deane's Doctrine of Naval Architecture, 1670	
Part 1 Arithmetic	33
Part 2 Hull Design	51
Part 3 Rigging	81
Appendix	115
A Contract Specification	116
Report on the *Warspite*	120
A Manifest for Carved Work	121
The *Resolution* Drawings	122

DEANE'S DOCTRINE OF NAVAL ARCHITECTURE, 1670

INTRODUCTION

Deane's Career

Little is known for certain about the early life of Sir Anthony Deane, but it is generally believed that he was born in 1638 in Harwich, the son of a master mariner. In 1660, at the age of 22, he was Assistant Master Shipwright in Woolwich Dockyard, and his rapid rise suggests that he had been apprenticed to Christopher Pett, the Master Shipwright. In August 1662 he first met Samuel Pepys, the Clerks of the Acts and a member of the Navy Board, and this acquaintance was to shape the rest of his career. Pepys found him knowledgeable and intelligent, and singled him out as a man to counterbalance the influence of the Pett family, who dominated naval shipbuilding. His brashness offended many, including the Lord Chancellor, but in October 1664 he was appointed Master Shipwright of Harwich Dockyard on the recommendation of Pepys. Thus began his active career as a ship designer.

At first Pepys was nervous that his protégé might fail him, but his first two ships, the *Rupert*, a 3rd Rate of 64 guns, and the *Fanfan* of 4 guns, were both considered highly successful. Deane built several other ships at Harwich, including another 3rd Rate, the *Resolution*. He was given a commission in the militia which was formed to resist a Dutch invasion, and was usually known as Captain Deane until he was knighted in 1673.

Harwich Yard was closed in 1668, after the end of the Second Dutch War, and Deane was appointed Master Shipwright at Portsmouth. This gave him the opportunity to gain experience of larger ships, and his first three-decker was the *Royal James* of 1671. He also indulged in several experiments, with a type of cannon called 'Punchinello', and with lead sheathing, which was first tried on the *Phoenix* of 1671.

In 1672 he was promoted to Commissioner at Portsmouth and thus became a member of the Navy Board. This removed him from direct control of shipbuilding at Portsmouth, but the standards of the time did not debar him from building Royal ships by contract, although he was a government

Left: Sir Anthony Deane. *(National Maritime Museum)*

official. He built several ships as a private shipbuilder at Harwich, including the *Harwich*, *Swiftsure*, and *Sapphire*, and the *Charles Yacht* at Rotherhithe.

In 1673 Deane became a member of the central administration of the navy, when he was appointed Controller of the Victualling Accounts. The duties attached to this office were light, and it was clear that he was intended to work with the Surveyor of the Navy, Sir John Tippets, in supervising the design, construction and repair of warships. By this time he had risen to a position of some social prestige. He was knighted in 1673, became Mayor of Harwich in 1676, and its Member of Parliament in 1678. His ships, particularly the yachts, had attracted the notice of the King and his brother the Duke of York, the future James II. In 1674-5, he built two yachts for Louis XIV of France at the orders of King Charles. He took the yachts to France personally, and was rewarded by Louis.

In 1677 an act was passed by Parliament for building thirty new ships of the 1st, 2nd, and 3rd Rates, and Deane, with Tippets, was given responsibility for deciding the dimensions of the ships. These dimensions were to provide the basis for development over the next seventy years. Although they were increased gradually over the period and the system of establishments of dimensions soon became over-rigid, the dimensions of 1677 represented the first attempt at standardisation of design, and at the time they were considered a great advance. Deane's most time-consuming work for the 1677 programme was in finding timber to build the ships. This was in short supply because of the after effects of the Civil War, the three Dutch Wars, and the Great Fire of London. Deane toured the country in search of suitable materials, and he was successful in that the ships were completed (though not on time). His opponents later claimed that the use of poor quality timber had caused the ships' early decay, but Deane successfully defended himself against this charge.

Deane's position still largely depended on the support of Pepys, now Secretary of the Admiralty, and Pepys in turn depended on the King and the Duke of York, who, despite his enforced resignation as Lord High Admiral in 1673, was still a considerable influence behind the scenes. By 1678 the campaign against the Roman Catholic Duke, inspired by the 'Popish Plot', was building up, and Pepys and Deane were soon caught in the Opposition attack. Deane was especially suspect because of his work for the King of France, and in 1679 he and Pepys were imprisoned on a charge of treason. They were later freed on bail, and eventually cleared in 1680, but they had both been forced out of office. For the next few years Deane appears to have made a good living as a private shipbuilder, though nothing is known about the ships he built.

By 1685 the King had regained control of the political scene. Pepys was re-appointed Secretary of the Admiralty in 1684. It was clear that something had to be done about the condition of the fleet, which had been severely neglected during his absence, and it was decided to set up a Special Commission to supersede the old Navy Board and undertake the work of repair. Pepys went to great lengths to ensure that Deane had a place on the commission on his own terms. Deane demanded a salary of £1000, twice what the King had proposed, as compensation for his loss of earnings as a private shipbuilder. Pepys ensured that he got it, by presenting James II, who had recently succeeded on the death of his brother, with a list of the best known ship-

Right: The Navy Office, as rebuilt after the fire of 1673.
Below: Samuel Pepys. *(National Maritime Museum)*

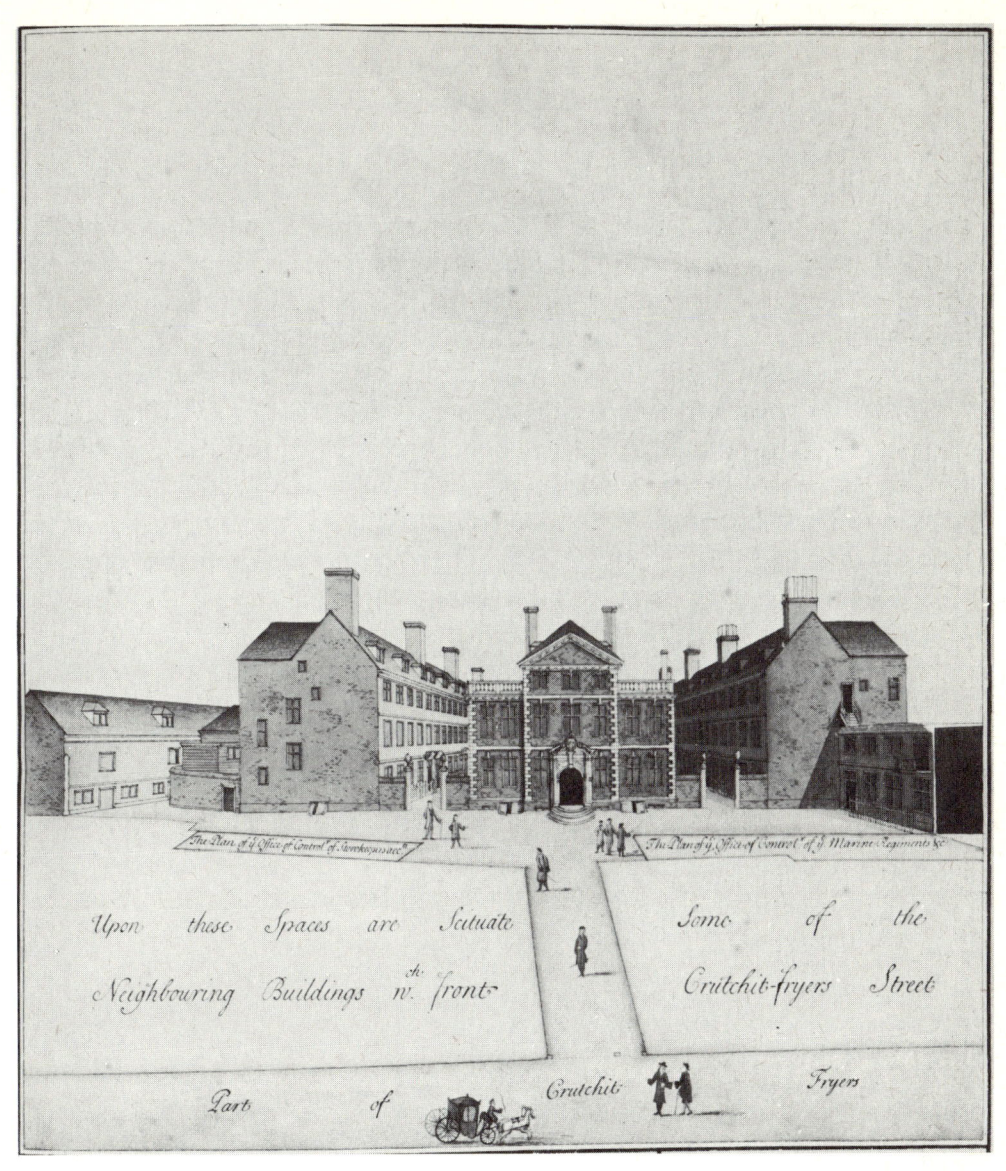

wrights in the country, and the reasons why every one of them, except Deane, would be unsuitable. Between 1686 and its dissolution in 1688 the Special Commission repaired and rebuilt a total of 66 ships and vessels, and built four new 4th Rates. As a result the King was able to send out a strong fleet in 1688 to oppose the invasion of William of Orange.

Unfortunately for Deane and Pepys, the fleet failed to stop William, and King James fled. Deane and Pepys were too closely associated with the ex-King to remain in office, and neither was ever employed in a government post again. They were imprisoned once more, under suspicion of plotting to restore King James, but they were released after a few months. They had to defend themselves against charges that the Special Commission had been corrupt and wasteful, which they did successfully.

After this Deane returned to obscurity. He appears to have retired, though he is said to have instructed Peter the Great in shipbuilding in 1698, and his son went to live in Russia. He died in 1721, at the age of 83.

Deane's Ships

Deane's career as an active shipbuilder for the Royal Navy lasted only ten years, but during that time he built 25 ships and vessels, out of a total of 94 built. It was an important period in the development of the warship, when the concept of the specialised ship of the line emerged, and developed into a form which was to last, with some modifications, until the end of the age of sail. It was also a period of conflict between two distinct concepts of warship design.

The early Stuart Kings, James I and Charles I, had preferred larger ships — either three-deckers like the *Prince Royal* and the *Sovereign of the Seas*, or high-charged 'great ships', with high and strong forecastles, intended to resist boarders. The leaders of the Commonwealth, on the other hand, had favoured smaller, faster ships. They built large numbers of one- and two-decked ships, partly based on the Dunkirk frigates and on the *Constant Warwick*, said to have been the first English frigate.

They also developed the line of battle, a tactical formation which was to survive as long as gun-power remained predominant. Before the First Dutch War of 1652-4 there had been no great fleet action since the Armada campaign of 1588, when gun-power first proved its effectiveness at sea. In the years after 1588 it was assumed that any future battle would consist of engagements between single ships and small squadrons, attacking, firing one broadside, turning round to fire the other, and then, if the enemy was sufficiently weak, boarding him. The Commonwealth Generals-at-Sea, experienced military men before they took command of the fleet, attempted to impose some order on the formation, and decreed that the fleet should form a single line ahead, thus deploying its great gun-power to the best advantage. It was to be many years before the line of battle became as rigid in practice as it was in theory, and battles often descended into mêlées after a time, but the concept was beginning to have an effect on ship design by 1665, when the Second Dutch War began, and Deane started his first ship.

At the end of 1664 it was decided to build four ships to the dimensions of the *Mary* (ex-*Speaker*) of 60 guns, one of the largest and most succesful of the Commonwealth's 'frigates' (at that time usually denoting a two-decker ship). Deane was to build one, along with Peter Pett of Chatham Dockyard, Shish of the Deptford yard, and Sir Henry Johnson, a private shipbuilder of Blackwall. A few weeks later Deane was ordered to build another, and a contract was signed with William Castle of Deptford for a sixth.

Originally they were all to be 116 feet on the keel and 34 feet 6 inches broad, but soon these dimensions were changed. Castle had his contract amended, with the approval of the Navy Board, so that his ship would be 36 feet 6 inches broad, and Pepys suggested that Deane should build his to similar dimensions. It seems that the King also intervened personally at some

The great ship and the frigate:
Left: The *Constant Reformation* as drawn by Van de Velde the Elder in 1648. *(National Maritime Museum)*
Below: A model of a frigate of *c*1650. *(Science Museum)*

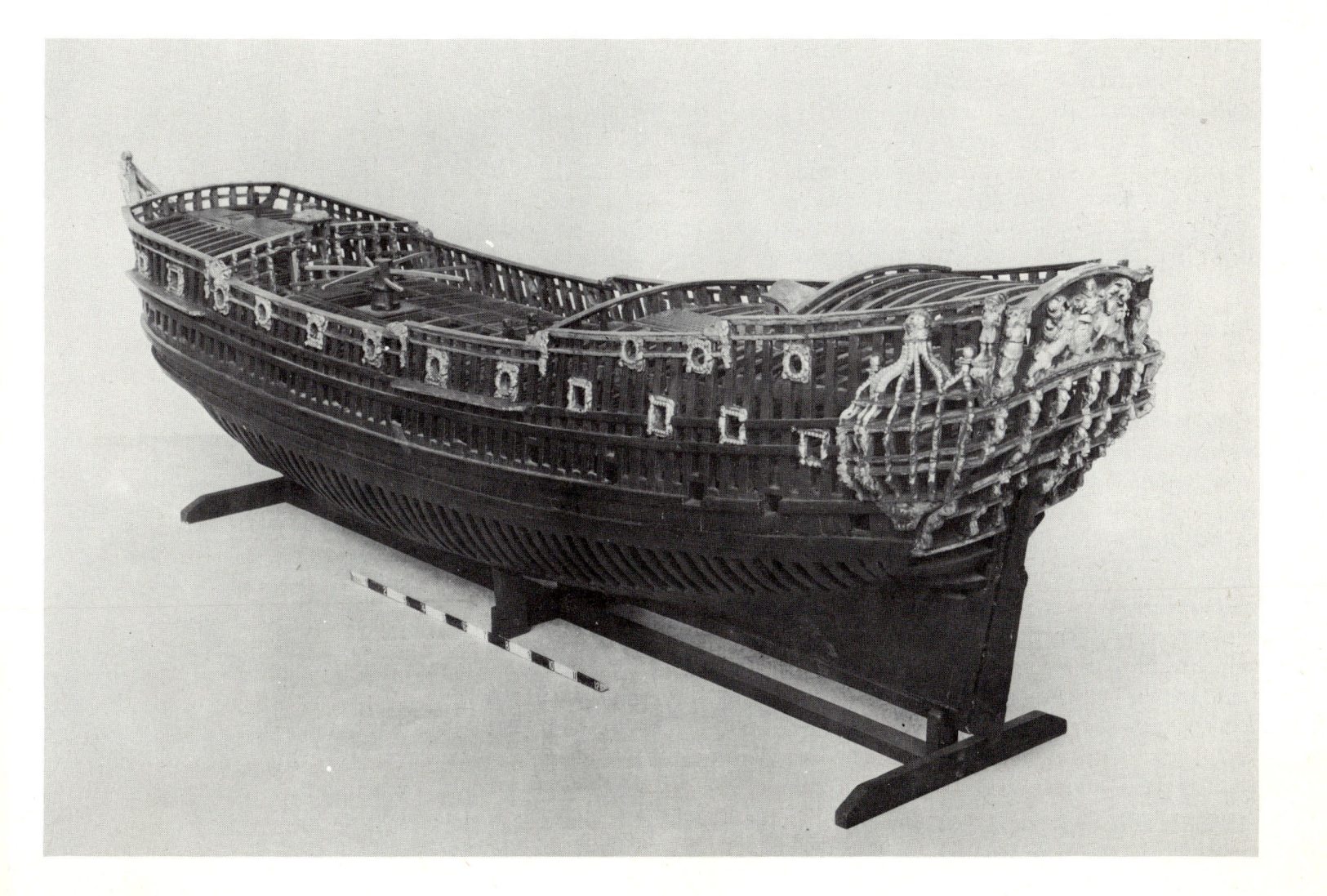

stage, and directed Johnson and Castle to increase the breadths of their ships, possibly because he was influenced by the design of several ships being built in Holland for the French Navy.

In any case the six ships all ended up considerably larger than originally intended, larger even than the amended proposals. Shish's ship, the *Cambridge*, was 121 feet on the keel when she should have been 116, and Johnson's *Warspite* was 38 feet broad. Deane's first ship, the *Rupert*, was the smallest of the six, but even she was 3 feet longer than intended. There are several possible explanations for this increase in size. One is that the private shipbuilders deliberately built larger because they were paid by the ton, and by increasing the dimensions slightly they could make more money. Another is that the standards of measurement were very low among most shipwrights. Shish of Deptford, according to Pepys, was unable to measure a piece of timber accurately. It is perhaps significant that Deane, a skilled draughtsman, came closest to the planned dimensions. Most important, the central administration had no real control over shipbuilding. It was unable to make up its mind what it wanted, and there was no recognised expert who could be easily consulted. The Navy Board was forced to rely on the advice of an outside body, Shipwright's Hall, and Pepys did not know whether to trust Deane, Pett, or Castle for advice.

Yet on the whole the new ships were remarkably successful. Partly by accident, partly through a series of insights, a new type of fighting ship had been evolved. The large two-decker was found to be the ideal ship of the line. It was much cheaper than the three-decker in proportion to its size, and with skilful design it could be made to sail almost as well as a smaller ship. It was strong and powerful enough to hold its place in the line of battle against the larger ships. In the 1690s diversions were made into ships of 50, 60, and 80 guns, but the ship of 64 to 74 guns remained the backbone of the line of battle until well into the nineteenth century, and the 64s and 74s which made up the great bulk of the British battle fleet during the Napoleonic Wars were directly descended from the six 3rd Rates of the Second Dutch War.

Development continued over the next few years. Two larger ships, the *Royal Oak* and *Edgar*, were built. Already it was becoming clear that ships needed to be made broader. At the Four Days Battle in 1666 the English ships found that they could not open their lower gun-ports in a heavy sea, and thus their heaviest armament was put out of action. The French and Dutch were building broader ships, which gave more room to work the guns, were more stable, and could carry more provisions, in return perhaps for a slight loss of speed. This trend was reflected in the *Edgar* and *Royal Oak*, which were 38 feet 10 and 40 feet 6 broad respectively.

Deane, however, remained conservative in this respect. In 1671 he proposed to build two more 3rd Rates, which were to be 120 feet on the keel and 37 feet 6 inches broad — very similar to the dimensions of his second ship, the *Resolution* of 1667. The proposed ships would have been smaller than any of the 3rd Rates built since 1664, apart from the *Rupert* and *Resolution*, and much smaller than the *Royal Oak* and *Edgar*. They would have been narrow in proportion to their breadth, whereas many people, including the King and most of the Master Shipwrights, had already seen that increased breadth was a great advantage in a ship of the line. Deane was still rooted in the Commonwealth tradition of fast frigates, but speed was a second-

The line of battle in practice: the Battle of Schoonveld, 1673. *(National Maritime Museum)*

ary advantage in a ship of the line. Pepys claimed that before 1673 the English shipwrights had 'not well considered it that breadth alone will make a stiff ship', but his friend Deane was the main offender in this respect, for Johnson, Castle, Shish and Francis Baylie of Bristol had already built ships which were broader than Deane's.

Two things happened during 1672-3 to change Deane's mind. One was the failure of his new 100-gun ship, the *Royal Charles*. She was Prince Rupert's flagship in May 1673, but he changed out of her because she proved to be unstable, and she had to be girdled to increase her breadth. The other was the opportunity to look more closely at French design — especially the *Superbe*. France was an ally during the Third Dutch War, and a squadron of French ships was sent to support the English fleet. King Charles visited them in May 1672, and was particularly impressed with the *Superbe*, which according to Pepys 'was 40 feet broad, carried 74 guns and 6 months provisions; and but $2\frac{1}{2}$ decks'. Deane was ordered to find out more about her, and to build his two new ships as close as possible to her design. His opportunity came after the Battle of Solebay in the same month, when he was put in charge of repairs to the *Superbe*. He reported 'I am silent in my observations and shall not commit it to view'.

The two new ships were therefore made broader, though neither came up to the full 40 feet of the *Superbe*. The *Swiftsure* was the first, launched in 1673, and she was 38 feet 3 inches broad and 123 feet on the keel. The *Harwich*, launched in 1674, was 38 feet 10 broad, with a keel length of 123 feet 8 inches. Pepys was pleased with the success of the *Harwich*, and reported in 1675 'the *Harwich* carries the bell of the whole fleet, great and small'.

In 1677, when Deane was given part of the responsibility for deciding the dimensions of the new ships, he had clearly been converted to the idea of greater breadth. Pepys claimed that the proportions of the ships of 1677 were based on those of the *Harwich*, but in fact the 3rd Rates were larger, and broader in proportion. Many people had contributed to the development of the 70-gun ship. The example of the Dutch and French had inspired the building of bigger and broader ships; the King had intervened several times to increase the size of his ships; Johnson had built the *Warspite* in 1666 to proportions remarkably similar to those settled in 1677; and Baylie's *Edgar*, and Shish's *Royal Oak*, had shown that larger ships were practical. Pett had designed the *Speaker* which had provided the basis for the original design, and he may have been instrumental in increasing the dimensions in 1665. Deane's contribution was far from negligible; he had proved with the *Harwich* that 70-gun ships need not be sluggish. It can never be proved whether he or Tippets played the major role in the dimensions of 1677, but it should be remembered that by that time Deane had built four 3rd Rates of the new type, more than any other builder, and Tippets none. The design of the ships of 1677 was to have a profound effect. The form of the 70-gun 3rd Rates was finalised, and lasted until they began to be replaced by 74s in 1755. The dimensions were increased slowly over the years, and more rapidly between 1745 and 1755, but the dimensions and layout of 1677 provided a solid basis. It is a great exaggeration to say, as Pepys seems to suggest, that Deane and the French influence alone were responsible for the development of the 70-gun ship, but Deane played a full part in the evolution of the large two-decker, the most consistently successful type of sailing warship between 1660 and 1820.

Deane also built three 1st Rate ships of 100 guns, but there is no evidence that he made any lasting contribution to the design of the larger ships, and the individual ships themselves were not particularly successful. The first, the *Royal James*, was lost at the Battle of Solebay in 1672, when she was less than a year old. The second was the *Royal Charles* which proved unstable in 1673 and was rarely used again until she was rebuilt as the *Queen* in 1693. The third, also called the *Royal James*, had no obvious flaws but achieved no great distinction.

Although Deane's *Doctrine* concentrates on the design of a 3rd Rate ship of the line, it is clear that his heart was not in the type of sturdy, stable battleship that was eventually to evolve. Deane, who had served his apprenticeship in the days when the Commonwealth had built fast frigates, was always attracted by speed, and in 1690 he claimed that the building of fleets of large ships had been a mistake — fast frigates and fireships, properly handled, could destroy them all. His career began with a spectacular success by one of his smaller ships. The *Fanfan*, a tiny 6th Rate of 4 guns, was launched in 1666 and joined the fleet in July of that year, just before the St James' Day Battle. Prince Rupert sent her to bombard the Dutch flagship,

Top right: Deane's *Resolution* of 1667, by Van de Velde. *(National Maritime Museum)*
Bottom right: The *Superbe*, also a Van de Velde drawing. *(National Maritime Museum)*

15

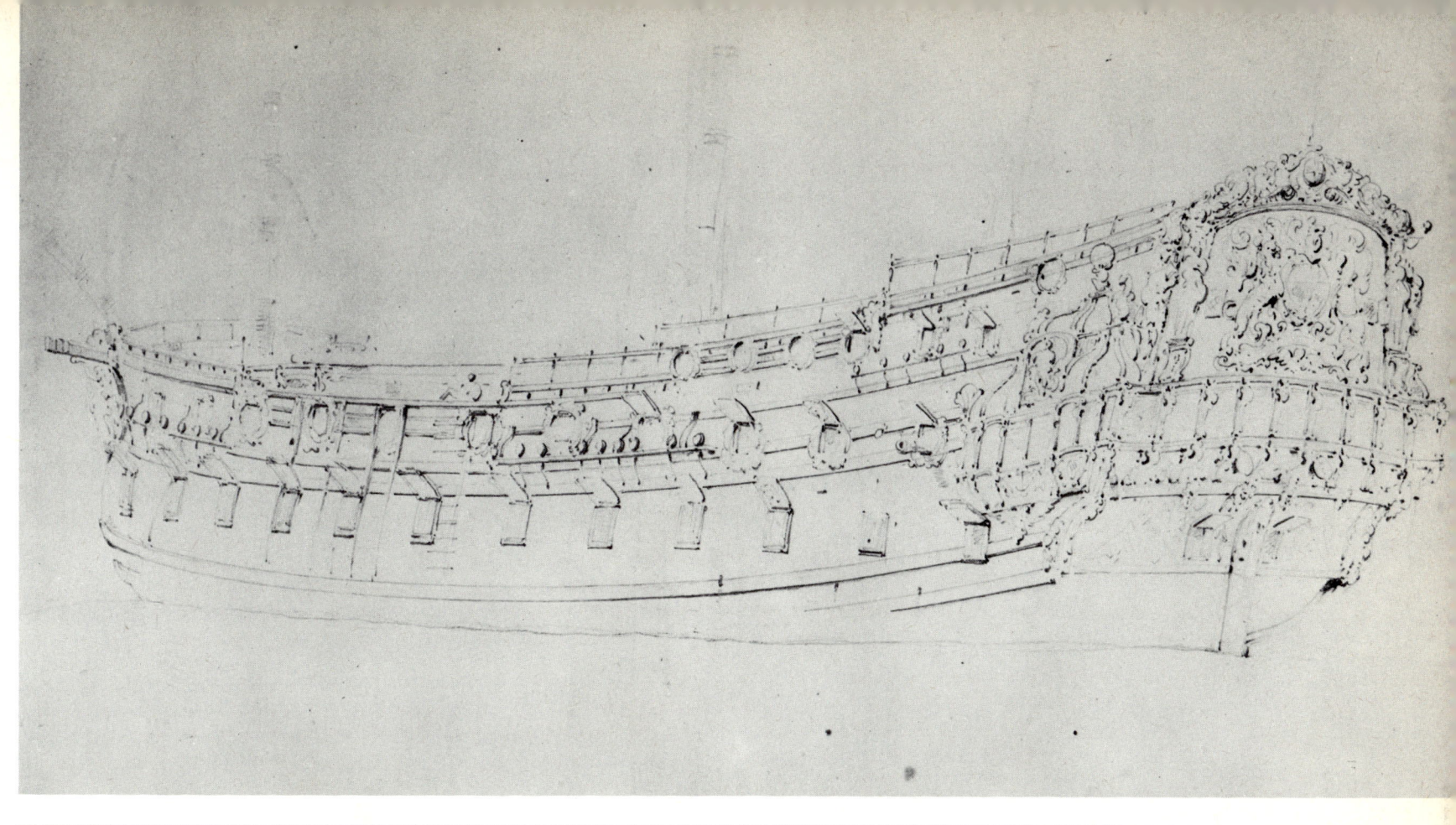

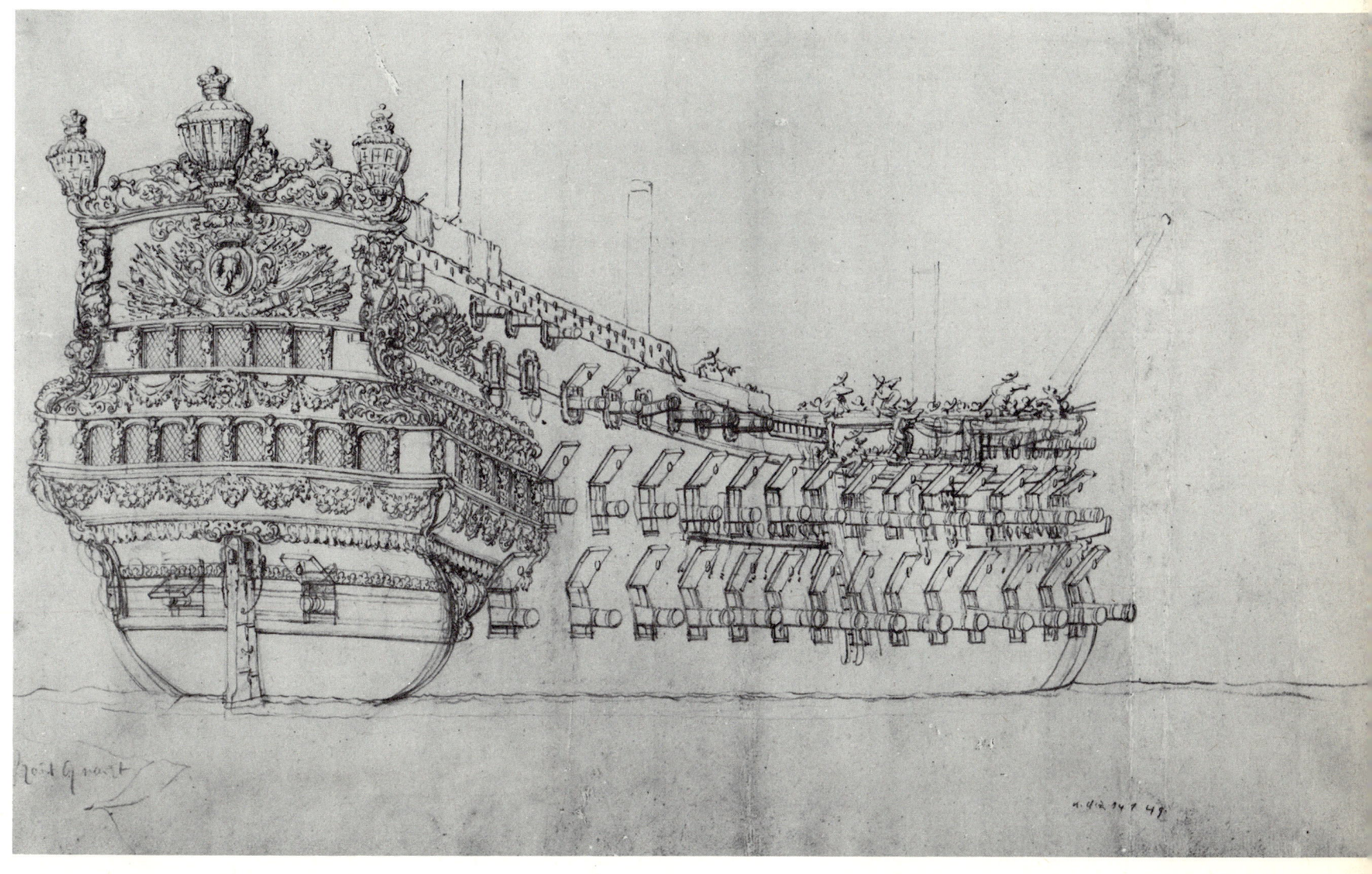

while keeping out of the way of the return fire. This she did, 'to the amusement of the English', and verses were composed by one 'IS', 'in honour of the *Fanfan*, which routed Van Tromp and the Belgic Navy'.

Perhaps his most successful small warship was the *Greyhound* of 1672. Very often we rely solely on Pepys for good reports of Deane's ships, but in this case praise came from several different quarters. After her launch it was reported that 'Never was a stiffer ship of that burden. She steers singularly well, keeps a weather helm, and though there was a great sea she never missed staying ... We believe she will be as good a sailer as was ever built in England, and as good conditions for accomodation, far exceeding any of the yachts.' Sir Henry Sheres chose her as an example of a fast English frigate, and said she was 'built with much art and judgement'. In 1673 she was mistaken for a privateer because of her good sailing qualities, and after the Third Dutch War she was employed as a royal yacht because of her good sailing qualities and accommodation.

The quality of speed made Deane's yachts particularly attractive to the King and his brother, who were keen amateur sailors. The first, the *Cleveland* of 1671, was found to be an hour faster than any of the other yachts over a four hour course, and the *Charles* of 1675 was even faster. Because of this Deane was eventually given the task of building two yachts for the French King, an event which had unfortunate repercussions later.

Deane also built two experimental ships. The *Nonsuch* was built by Deane according to the proposals of Louis Van Heemskirk, a renegade Dutchman, who claimed that she would sail half as fast again as any of the ships in the navy. She appears to have been considered succcesful, and her inventor was promised payment for his work, but there is no sign of her design being copied, and no indication of what made her unique, except that in later years Deane and Pepys remembered 'the ridiculous advice we took to

Above: The *Royal James* of 1675.
Top right: A model of a 1st Rate. The triple lower wales are highly unusual, and outside Deane's *Doctrine*, this is the only known example of their use. This model may be a preliminary design for the *Royal James* of 1671.
Bottom right: The *Charles* yacht. *(All National Maritime Museum)*

INTRODUCTION

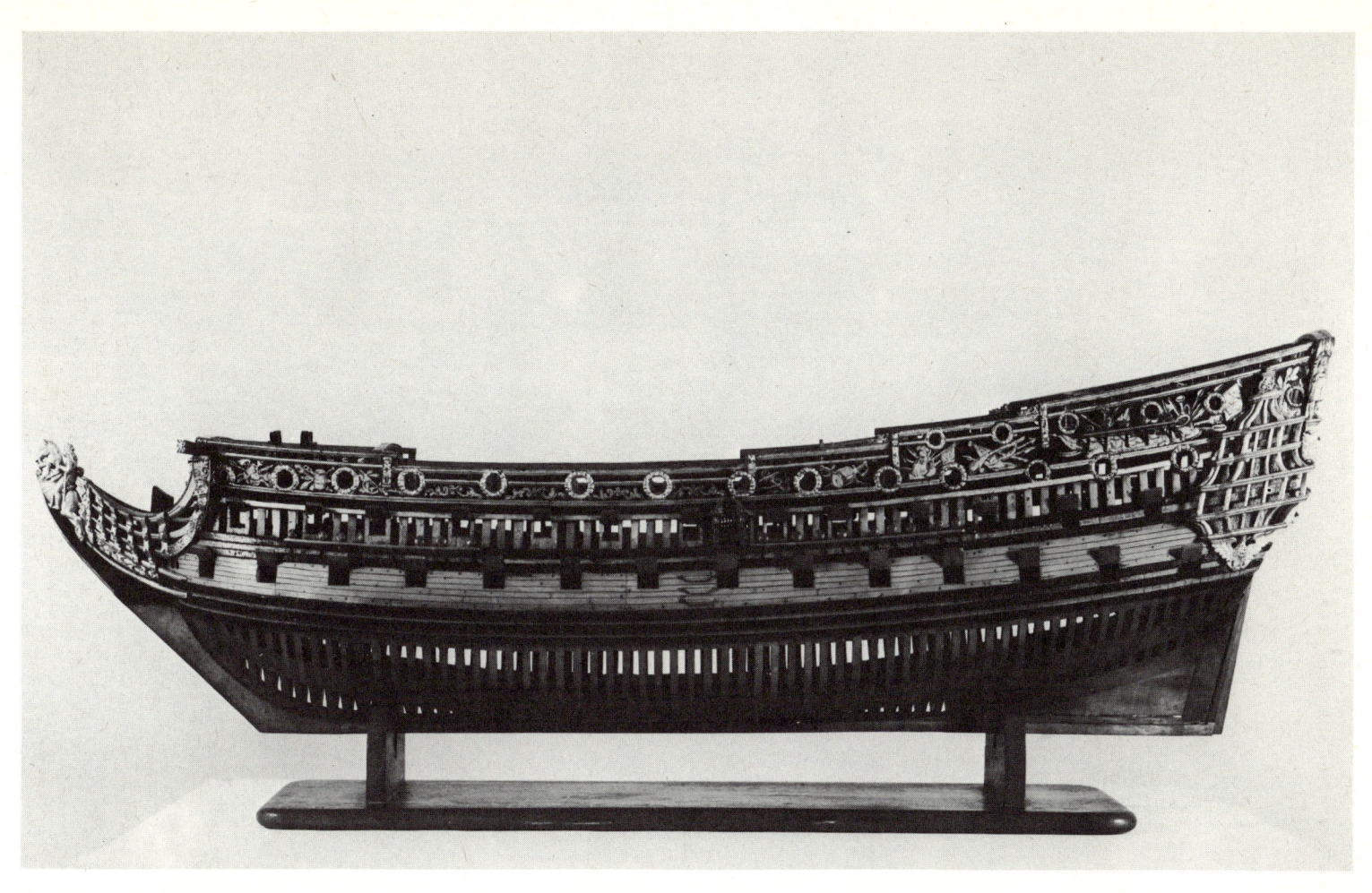

DEANE'S DOCTRINE OF NAVAL ARCHITECTURE, 1670

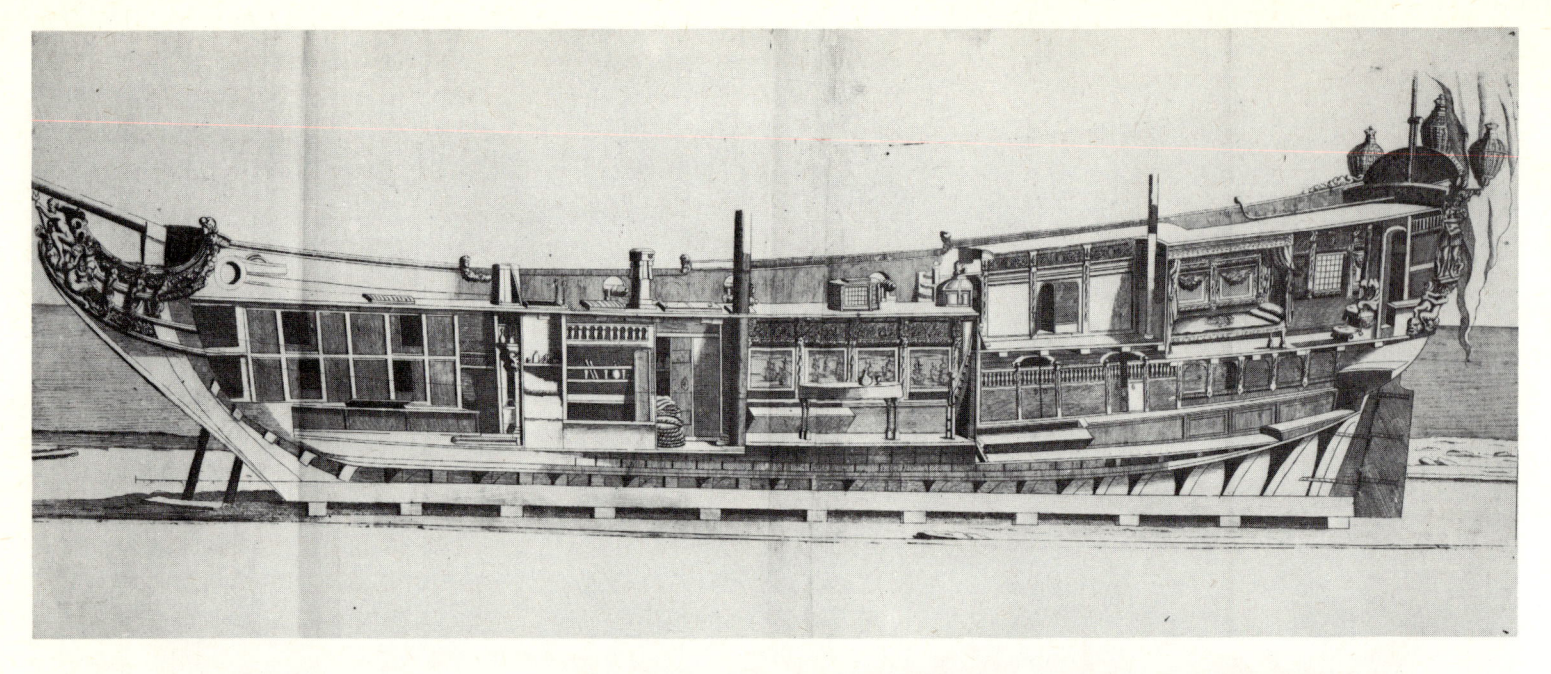

DEANE'S DOCTRINE OF NAVAL ARCHITECTURE, 1670

18

Top left: A cross section of a Stuart royal yacht.
Bottom left: A 20-gun 6th Rate of *c*1720, showing the sweeps in use. *(Both National Maritime Museum)*

build a ship by the Dutchman's invention with the grain of the timber and roots set all one way'. The second experiment was with lead sheathing. This was intended to prevent shipworm, which could rapidly destroy the underwater timbers of a ship in a warm climate. The first ship to be sheathed in this way was the *Phoenix* of 1671, built by Deane at Portsmouth. Deane played an important part in the experiments with lead sheathing, but ultimately they were a failure, because the electrolytic action of the seawater on the lead caused rapid decay. Almost a century later the problem was solved by using copper, with copper bolts instead of iron.

Deane's ships were considered fast by Northern European standards, but in the Mediterranean British ships could not compete with those of the Barbary corsairs. This was becoming a considerable problem, for Charles II found it necessary to mount several expeditions against them. Sir Thomas Allin commanded one in 1669-70, with Deane's *Resolution* as his flagship, and he reported 'My ships can do no good upon them [the Algerians] by sailing, as they will not fight unless by accident or force, which makes us all very weary of this type of warfaring.' The answer was to develop a specialised type of ship, for, as Pepys observed, 'Our seas require stronger and therefore heavier ships, which spoils their sailing, and therefore not the best in calm seas, and therefore we should have ships built to be employed only in the Straits' (ie, the Mediterranean).

The new type, called the galley-frigate, was largely copied from the French, and Deane was mainly responsible, with the help of his son, for its introduction to the Royal Navy. The new ships were intended to find the best compromise between oars and sails, and they had very fine lines for their times. They carried a few guns at each end of the lower deck, but most of the space there was devoted to a tier of sweep-ports, to be used for rowing the ship in a calm, and most of the guns were carried on the upper deck. The first two were the *Charles Galley* and the *James Galley*, launched in 1676, and built to a draught prepared by Deane's son, Anthony junior. They were regarded as very good sailers, though slow under oars. They could be seen as a step towards the classic eighteenth century frigate, which had a completely unarmed lower deck, and also a step towards specialisation. In the *Doctrine* Deane suggests that ships of all sizes should have similar proportions, but it was now becoming accepted, perhaps for the first time, that the ship of the line should be broad and stable, and the small cruising ship narrow and fast.

Ironically, Deane's work with the small fast ships had much less lasting effect than his contribution to the development of the ship of the line. The galley-frigate was ultimately abandoned, and the ships were replaced by ordinary 20- and 40-gun ships, though many of the 20s still carried sweep-ports, as did most sloops and frigates in the second half of the eighteenth century. There was much less activity against the Barbary corsairs in the eighteenth century, and the galley-frigates suffered a high loss rate when transferred to the North Sea and the Atlantic. Royal yachts also declined in importance in the eighteenth century. None of the succeeding monarchs shared Charles' and James' love of sailing, and they tended to prefer slower, more stable, and more comfortable vessels. Deane's designs may have had some indirect influence on the 20-gun ships which were built to fight the French privateers in the early 1700s, but in general the early eighteenth century had no great interest in speed, and bureaucracy and economy replaced experiment and ingenuity in ship design. The benefits of specialisation were forgotten, and by 1719 ships from 100 to 40 guns were built to almost exactly the same proportions. By the 1750s, when the 'true frigate' was developed, any trace of Deane's direct influence had long disappeared.

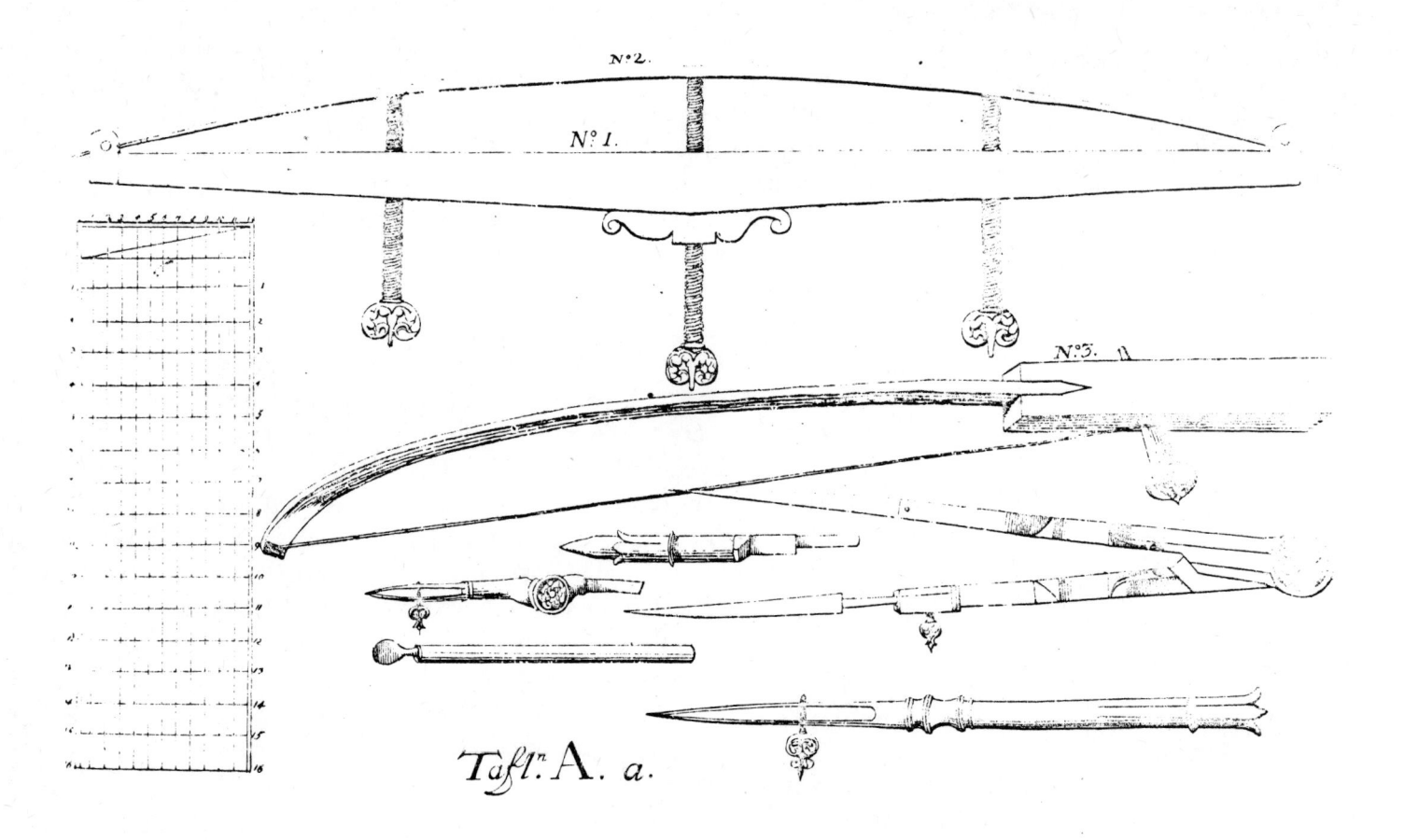

      Deane must have built a number of merchant ships during the years 1680-4, to earn his fees of £1000 a year as a private shipbuilder, but nothing is known about them. The main demand at this time was for ships which could carry a large cargo with a small crew, very different from Deane's typical designs for the Navy. Either he was particularly adaptable, or he specialised, perhaps, in private yachts, which would be very much in keeping with his reputation and experience.

      Probably Deane's reputation as a shipbuilder has been inflated by his association with Pepys, who was the most articulate man of his time, and had no scruples about advancing his friends and denigrating his enemies. We are given the impression that Deane was truly outstanding, far ahead of any of his contemporaries. This does not stand up to examination, for men like Shish, Johnson, Tippets, and especially Peter Pett, also made valuable contributions. Probably Deane was the best shipwright of his time, and it is likely that he made more contributions to the development of the sailing warship than anyone else of the period. But the competition was much closer than Pepys would allow.

A naval architect's tools in 1690. From top to bottom: an adjustable batten; a draughtsman's bow; compasses, with pencil and pen attachments; and a draughtsman's pen.

Deane's System

Deane's *Doctrine of Naval Architecture* was, according to the title page, 'written in the year 1670 at the instance of Samuel Pepys Esq.' Perhaps it developed out of the lessons in naval architecture which Deane gave to Pepys during 1663-4, which could account for its conversational style. On the other hand, perhaps the title page does not tell the whole story. There are many references to 'the young artist', and this could hardly have meant Pepys, who was then aged 37, and 5 years older than Deane. Deane also announces his intention to 'leave nothing unfolded which may advance anything to the meanest capacity'. This was not the way to flatter one's patron. It seems quite possible that Deane had once intended his doctrine for publication, and only later thought to present it to Pepys. In this he may have been influenced by the success of Edward Bushnell's work, *The Complete Shipwright*, which was already in its third edition by 1670.

In any case, Deane's work is one of the most important texts in the history of naval architecture. It is superior to Bushnell's, the only English contemporary writer on the subject, in that it gives much more information, although Bushnell is much more succinct. It is also superior to Bushnell's work in both the quality and quantity of the illustrations, though Bushnell may have been let down by his engraver. It was written by a practical ship designer, and we can test Deane's theories against the actual ships he built. It was written during a period of controversy in naval architecture, and Deane constantly makes us aware that there were those who differed from him on certain points, for example on the shape of the rising line. Often he dismisses his opponents with a few derogatory remarks, but in other cases he gives arguments for his own position, and this gives us much insight into the system of ship design. Above all, Deane's work is valuable because it gives us the earliest complete plan of an English ship. It is not a specific ship, but with the use of other information it is possible to reconstruct one of Deane's ships with a reasonable degree of accuracy.

By 1670 naval architecture was slowly progressing from a medieval craft into a modern science. In the past ship design had been considered part of the training of every shipwright, but by 1664 Bushnell was complaining that most were only trained as manual workers, and the secrets of design were known only to the favoured few. The use of draughts in shipbuilding was increasing, though not all shipwrights could use them. In 1665 the naval overseer at Bristol complained that 'there are no workmen there who understand the manner of doing it'. Shish, Master Shipwright in Deptford Dockyard, according to Pepys was illiterate, alcoholic, and unable to measure a piece of timber properly. Yet even Pepys had to admit that he had built some fine ships. In contrast Deane was one of the more modern school, working from the drawing board rather than by experience and feel.

It is often asserted that Deane was the founder of scientific shipbuilding, and this claim is worth examining. He provides a long section on arithmetic and geometry, but it is not true to say that his doctrine 'served as a pattern for the textbooks on the shipwright's trade' in that it 'starts with studies of arithmetic, mensuration, and geometry'. Bushnell, first published in 1664, had provided a similar section, and there are signs that Deane's style was influenced by it. Deane, like most writers on the subject up to the first half of the nineteenth century, gives an impressive number of calculations throughout his work, but these are purely geometric, giving a means of forming the shape of the ship. With the exception of the calculation of the draught of water, they give no indication of the likely performance. He also attempts to impose a series of proportions on his design. Thus the breadth is to be $\frac{3}{10}$ of the length of the keel, the floor $\frac{1}{3}$ of the breadth, the floor sweep $\frac{1}{4}$ of the breadth, and so on. This is, of course, not a truly scientific system, for it depends entirely on the experience or prejudices of the builder, and often it would have to be modified in practice. Deane, as we have seen, came to favour greater breadths in his later years. In any case, the breadth was normally decided by the owner of the ship, whether the government or a private shipowner, and most of the dimensions were laid down in the contract in considerable detail.

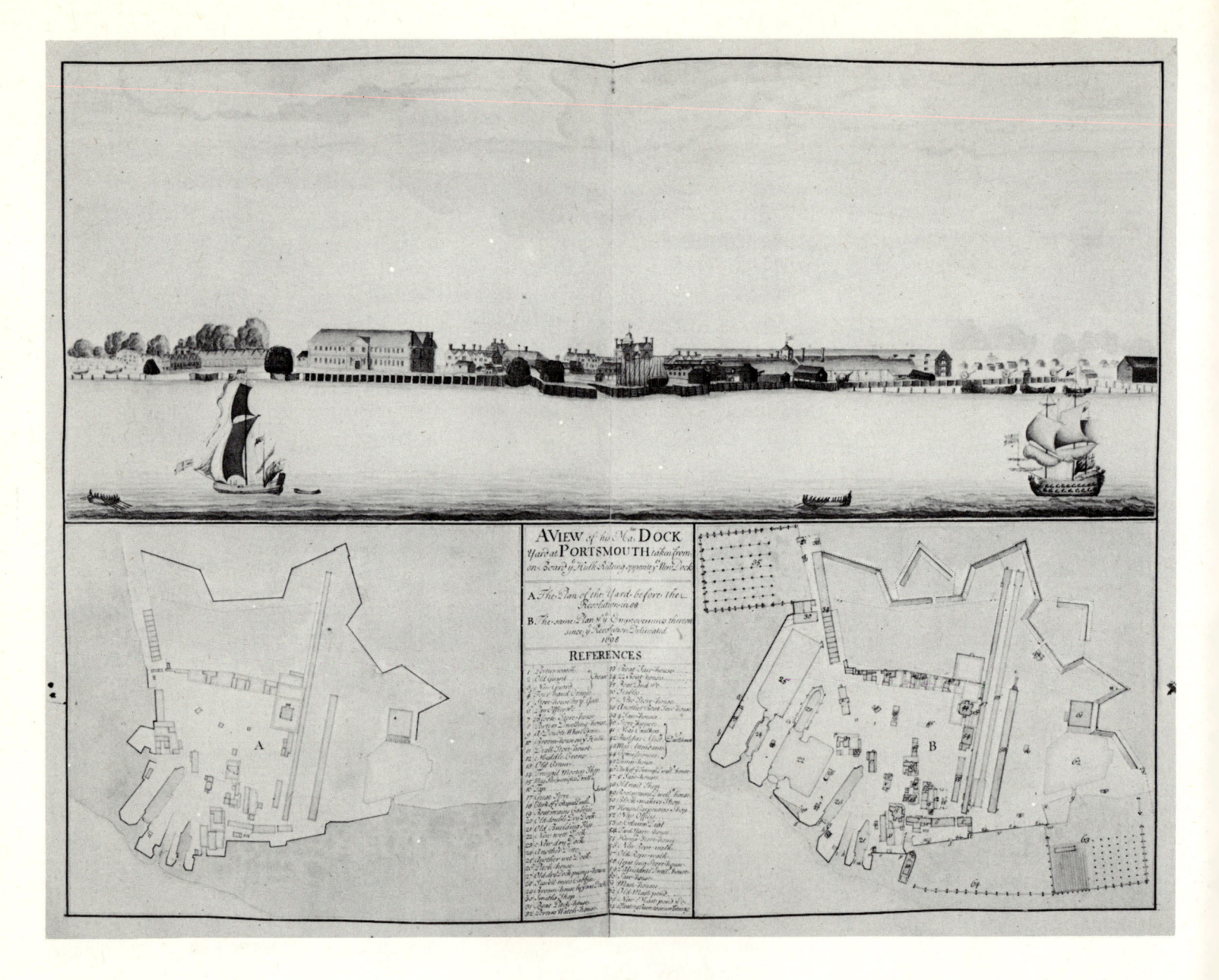

Deane's greatest claim as a scientific shipbuilder lies in the belief that he invented the system for calculating the displacement of a ship,

Portsmouth dockyard in Deane's time:
Above: A general plan.
Right: The dry dock and building slip.
(Both British Library)

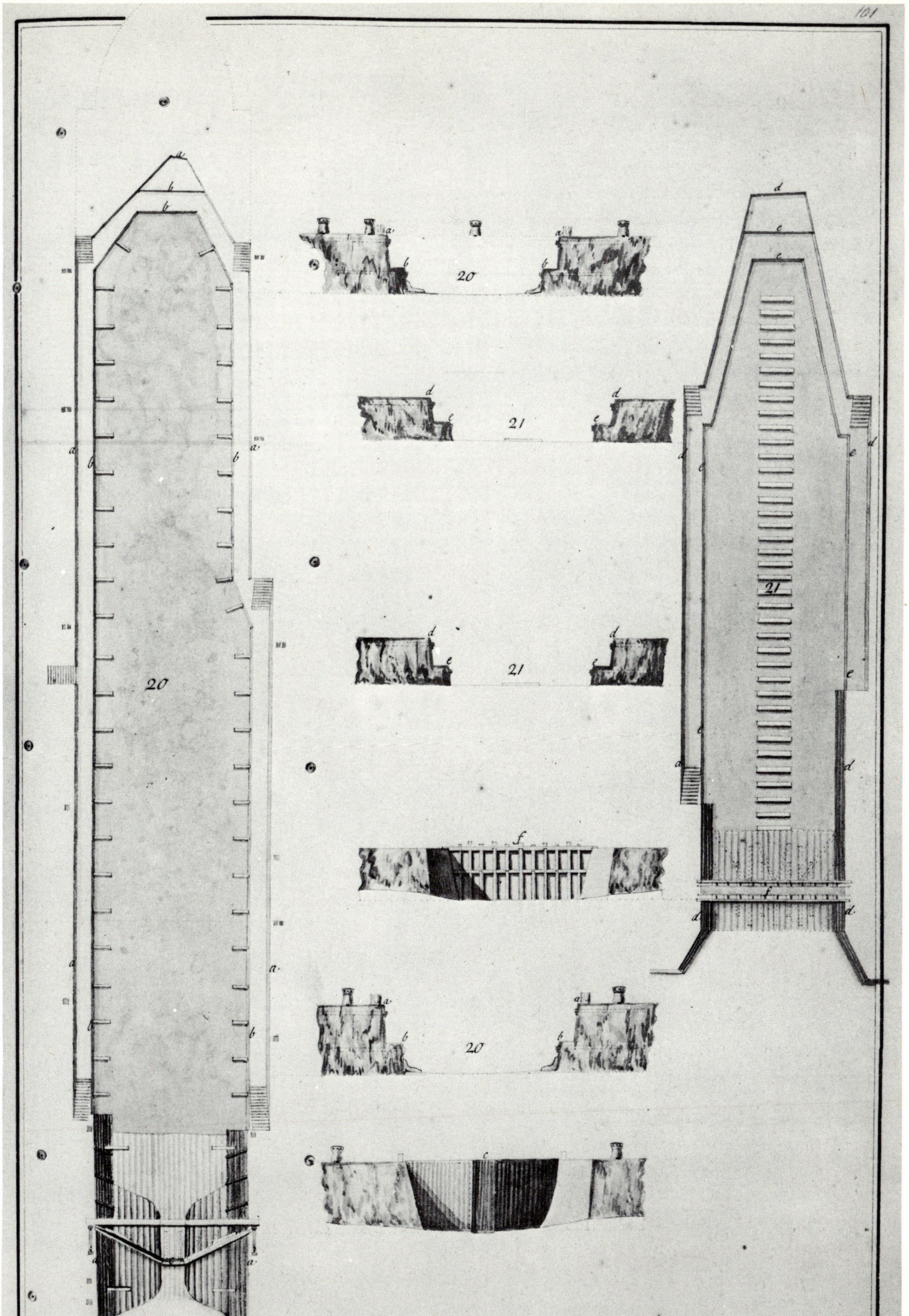
101
20
21
21
20
21

DEANE'S DOCTRINE OF NAVAL ARCHITECTURE, 1670

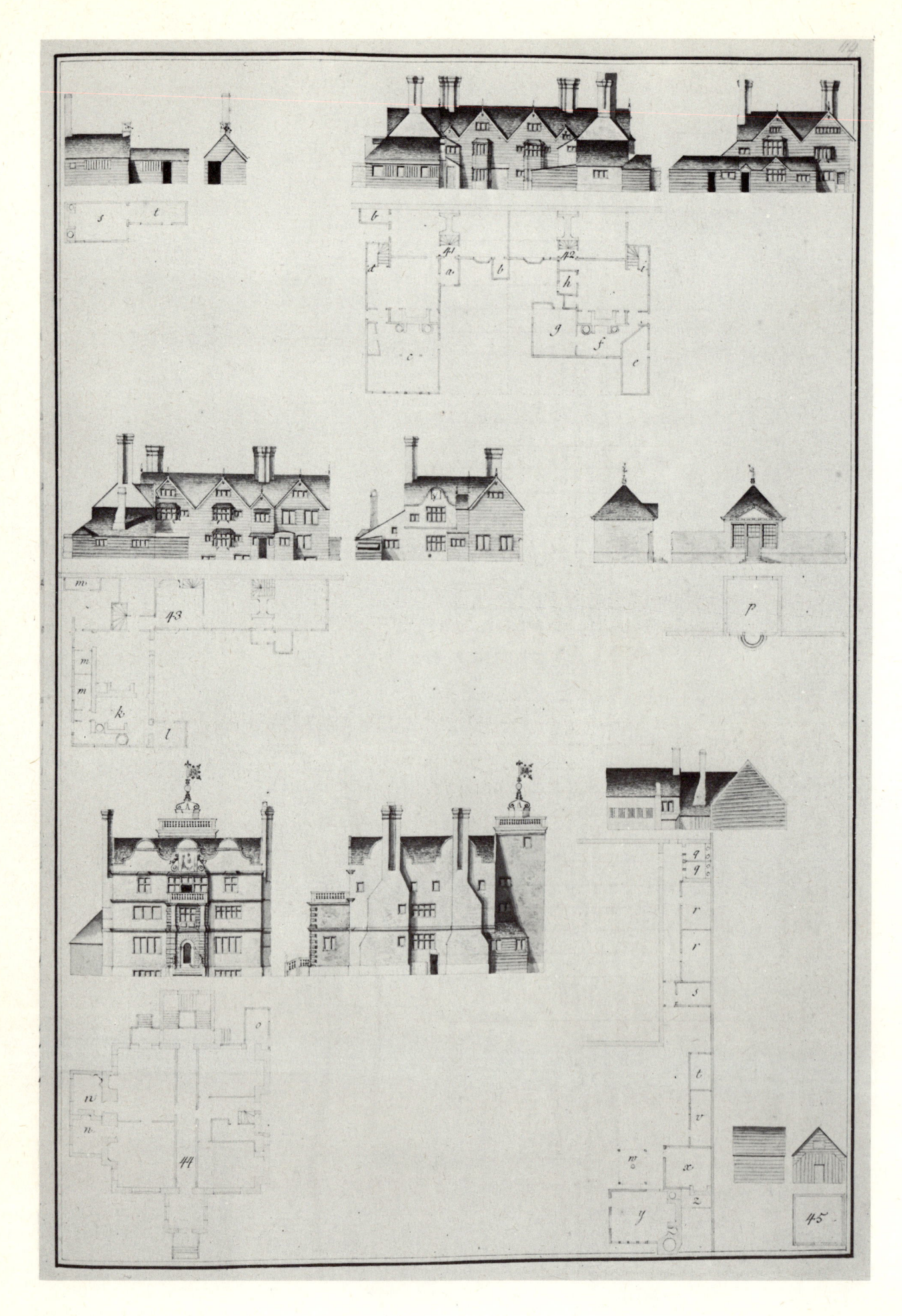

thus showing what her draught of water would be, and how high her gunports would lie from the water-line. Certainly this is a very important point. The system would avoid disasters like the *Royal Katherine* of 1664, whose lower ports were found on launch to be 'but 3 feet above water before she hath all her provisions and guns in'. It would help avoid the girdlings which were frequently necessary at the time because the original design did not work out in practice. Pepys was in no doubt that Deane had invented the method for calculating the displacement. 'He is the first that hath come to any certainty beforehand of foretelling the draught of water of a ship before she is launched,' he wrote in 1666, and he made several references to Deane's 'secret'. Yet Deane, despite his reputed arrogance, does not make this claim for himself in the *Doctrine* — indeed, he implies that the method was available to any shipwright if only he 'will take the pains, or his skill afford so much of that art'. It has been shown that as early as 1586 Matthew Baker was beginning to understand the principles involved, and was moving towards such a method. By 1678, at the latest, specially printed squared paper was in use for drawing the lines of ships, and this could provide an alternative method of finding the displacement. By 1685 Edward Battine, who was at best an amateur, knew of the existence of the method. If Deane had invented it in the mid-1660s, it must have come into common use surprisingly quickly for the times. There are no major works on English naval architecture between 1625 and 1664, so the question will probably never be proved either way, but the evidence suggests that the method had developed slowly over the years, and Pepys had allowed his enthusiasm for Deane to mislead him.

Deane spends much of his time describing how to draw the sheer draught of his ship. His methods appear to be typical of the times. Perhaps he was inclined to give a little more sheer to the wales than was normal, and this is a striking feature of many of the Van de Velde drawings of his ships. His method of drawing the shape of the stem, using a single curve, was also typical, but was soon to be superseded. The rake of the stem was being progressively reduced over the years, in order to give more support to the heavier armament being carried. From 1677 'straight stemmed' ships were built. Either the curve was no longer tangential to the keel, or it was greatly reduced in diameter and carried up to the gun deck by a straight line.

The most important and difficult task of a ship designer is to form the shape of the underwater body. This is obviously a very complex shape, and various techniques were used over the years to make it comprehensible. The earliest way was to use the same shape for all the frames, moving it inwards and upwards towards the bow and the stern, and joining it to the keel by means of a straight line or reverse curve. Probably this was the original meaning of 'whole-moulding', for a single 'mould' or template, could be used to cut out the whole frame. Certainly this was how the term was understood in the early nineteenth century, when it still survived in small boat design. A hull which is whole-moulded in this sense is easily recognisable, since the rising lines of the floor and breadth are parallel.

By 1586, at the latest, whole-moulding had taken on a different meaning. As in the old system, each frame was made up of several curves, and these had the same radii throughout the length of the ship, but their relationship to each other varied throughout the length. The two rising lines were no longer parallel. The rising line of the floor still rose rather steeply at the ends,

Left: The Commissioner's House, Portsmouth Dockyard in Deane's time. *(British Library)*

but that of the breadth was becoming increasingly flatter. This made the bow fuller, and more able to support a heavy armament. It also meant that the shape of each individual frame varied, though the curves which composed them remained the same.

This system, like its predecessor, had an inherent disadvantage. It was intended to form a shape which could be easily understood and reproduced by a shipwright, rather than an ideal one for moving through the water. Towards the bow and stern the large curves used in midships were unsuitable, and bumps tended to appear in the lines of the hull. Early shipwrights had no means of foretelling this, for they do not appear to have used any system of projection, but they probably removed the bumps in practice, by fitting battens to the side of the frame and trimming any surplus timber with an adze. Moreover, a draught drawn in this way does not show fully the ends of the ship, especially at the bow where the shape became particularly complex, and the curves changed rapidly. Builders had to interpret this important shape for themselves, and two builders working to the same draught would not necessarily produce identical ships.

By Deane's time this system was undergoing a change. The text of Deane's *Doctrine* suggests that he was using the second version of wholemoulding, and keeping the same curves along the length of the ship. In practice he was doing nothing of the kind. On his completed draught the radii of the floor sweeps forward are progressively reduced towards the bow, and aft they vary irregularly, but tend to reduce towards the stern. The breadth sweeps are progressively reduced towards the bow and stern. This gives a much fairer form than would be possible with the old system. Perhaps it is the real 'secret' of Deane's system. If so, it is not fully revealed in the *Doctrine*, for Deane does not even admit that he is reducing the sweeps, far less explain his method of doing so. Perhaps this was because he had no 'scientific' means of showing this, and relied solely on experience and judgement. Moreover, he does not explain how he draws the reverse curves which connect the upper parts of the frames to the keel, and these are rather roughly drawn on the completed draught.

Probably part of the answer is to be found in his use of water-lines. Deane mentions the drawing of water-lines at two feet intervals in order to prove the design, although he does not show them in any of his draughts. Deane's *Doctrine* is the earliest English document to mention water-lines in this way, although a draught of a 6th Rate which can be tentatively dated to around the same time also shows water-lines. It was an important step forward. By the early eighteenth century vertical and diagonal fairing lines had been added to the horizontal ones which Deane mentions, and it was possible to fair the hull on paper, rather than after construction.

In these respects Deane's *Doctrine* appears to be very advanced for its time. Yet, in other, more superficial respects, it is rather backward. The decorations shown on Deane's draughts are years out of date. All the draughts except one show ships with square decorations on the gun-ports, although these had been superseded by round decorations about ten years earlier. The beakheads of his ships are very low, perhaps to allow guns to fire over them. This was considered important in earlier times, but the line of battle had largely made it obsolete, and the old type of head was liable to damage in rough seas. The *Prince*, launched in the same year as the *Doctrine* was written,

Top right: The water-line of the *Greyhound* of 1672. This was used by Sir Henry Sheeres in an early form of test tank.
Bottom right: The lines of the *Berwick*, probably drawn by Edward Dummer in 1678. The squared paper was probably used for measuring displacement. *(British Library)*

INTRODUCTION

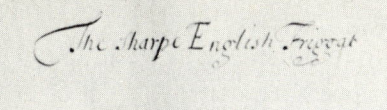

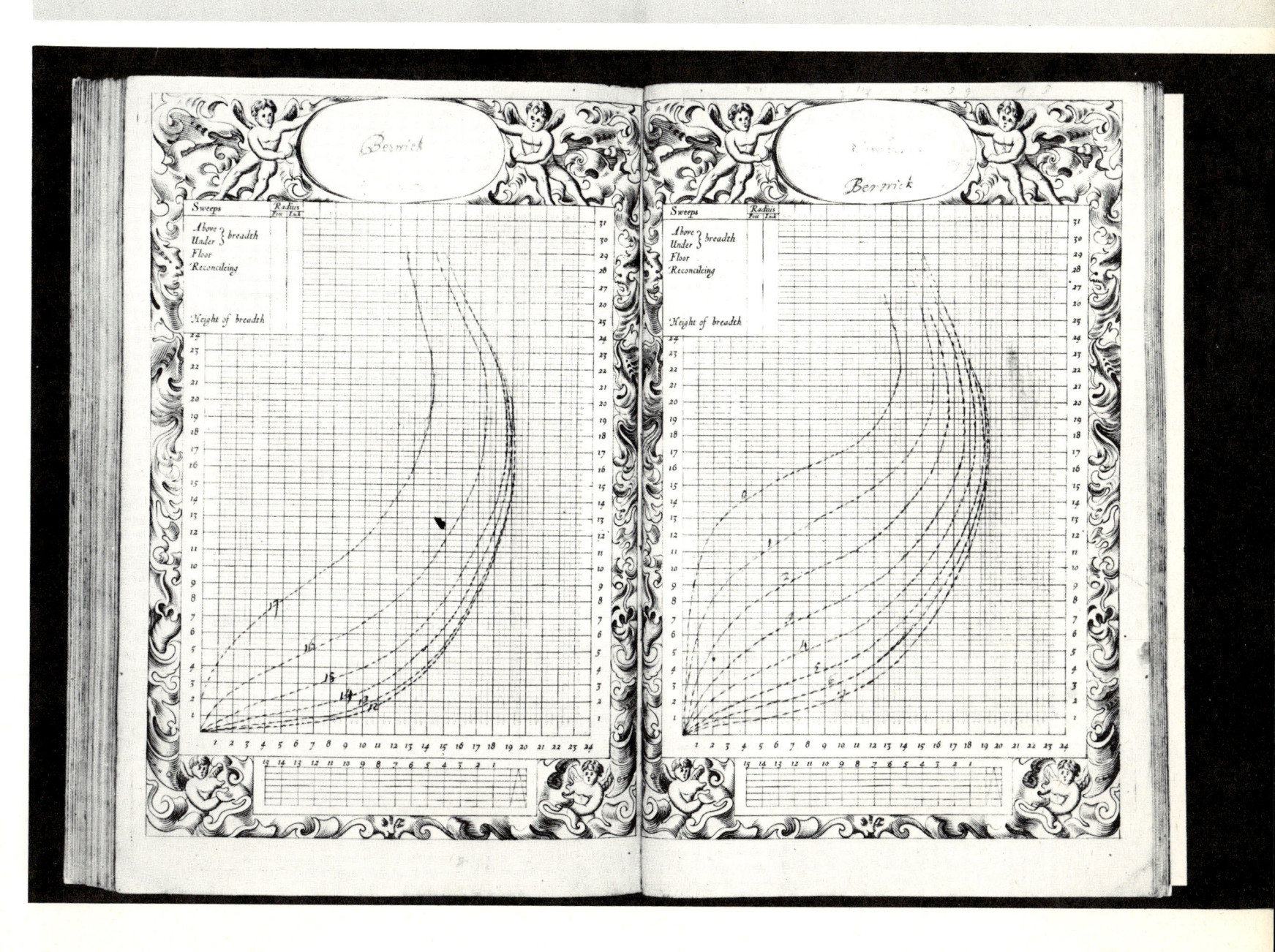

27

had a much higher beakhead than Deane's projected 1st Rate, and was able to carry forward firing guns only on the upper deck, whereas Deane's ship would have been able to carry them on the middle deck as well. The stern of the 3rd Rate is also unusual, in that it does not have the royal coat of arms which was invariably carried at that time, though it is apparently intended to show a royal ship, as the monogram 'CR' can be seen.

    These, however, are minor points. Information about the decoration of Charles II's ships can be found in Van de Velde drawings and in surviving ship models. The great value of Deane's *Doctrine* lies in the fact that it gives us more information than any other source about the exact hull form of English warships, and outlines the method by which they were designed. It also tells us a great deal about how far the practice of naval architecture had developed by 1670, and this is not to be found anywhere else.

SOURCES FOR SHIP LIST
Caution is necessary when dealing with the dimensions of ships of this period, and there are many variations between different lists. There were, for example, three different ways of measuring the keel, and possibly variations within these methods. These figures are largely based on Pepys' 'Register of Ships', but collated with other sources, including *Dimensions Book B* and the list of the Royal Navy published in *The Diary of Henry Teonge* (ed Mainwaring, 1927). The guns and men are those established in 1677 where applicable. Otherwise they are those given by Pepys.

Deane's Ships 1666-75.

Name	Guns	Men	Launched	Yard	Length on the Gundeck (ft ins)	Length of Keel (ft ins)	Breadth (ft ins)	Depth in Hold (ft ins)	Tons	Fate
1st Rates										
Royal James	102	800	1671	Portsmouth		136-0	45-0	18-5	1426	Burnt in the Battle of Solebay, 1672
Royal Charles	100	770	1673	Portsmouth		136-0	44-8	20-6	1443	Renamed Queen, 1693, and rebuilt
Royal James	100	770	1675	Portsmouth	163-1	132-0	45-0	18-4	1485	Renamed Victory, 1691. Rebuilt 1695
3rd Rates										
Rupert	68	395	1666	Harwich	144-1	119-0	36-3	15-6	832	Rebuilt 1703
Resolution	70	415	1667	Harwich	148-2	120-6	37-2	15-9	885	Rebuilt 1698. Lost 1703
Swiftsure	70	415	1673	Harwich	149-3	123-0	38-8	15-6	978	Rebuilt 1696
Harwich	70	415	1674	Harwich		123-8	38-10	15-6	993	Wrecked off Plymouth, 3 Sept 1691
5th Rates										
Nonsuch	40	175	1668	Portsmouth		88-3	27-8	10-10	359	Captured by French Privateer, 4 Jan 1695
Phoenix	40	175	1671	Portsmouth		90-2	28-6	11-2	368	Burnt to avoid capture by the French, 12 Apr 1692
Sapphire	30	130	1675	Harwich		86-0	27-0	11-0	346	Sunk at Newfoundland, 11 Sept 1696
6th Rates										
Fanfan	4	30	1666	Harwich		44-0	12-0	5-8	33	Made a pitchboat, 1693
Francis	16	65	1666	Harwich		66-0	20-0	9-2	140	Lost in hurricane in the Leeward Islands, 1 Aug 1684
Roebuck	16	75	1666	Harwich		64-0	19-6	9-10	129	Sold Dec 1683
Saudadoes	8	40	1670	Portsmouth		50-0	18-0	8-0	83	Rebuilt as 180 tons, 1673
Greyhound	16	65	1672	Portsmouth	93-0	75-0	21-6	9-0	184	Sold 13 May 1698
Lark	16	65	1675	Blackwall		74-0	22-6	9-2	199	Sold 3 May 1698
Sloops										
Spy	4	10	1666	Harwich		44-0	11-0	4-0	28	Sold Dec 1683
Prevention	4	10	1672	Portsmouth		60-0	12-0	5-0	46	Sold Dec 1683
Cutter	2	12	1673	Portsmouth		60-0	12-0	5-0	46	Cast away at Deal, Sept 1673
Hunter	4	10	1673	Portsmouth		60-0	12-0	5-0	46	Sold Dec 1683
Invention	4	10	1673	Portsmouth		44-0	11-0	5-0	28	Sold Dec 1683
Smack										
Swan	–	5	1666	Harwich		36-0	11-3	5-2	24	Taken by Dutch, Oct 1673
Yachts										
Cleveland	8	30	1671	Portsmouth		53-4	19-4	7-6	107	Sold 1716
Navy	8	30	1673	Portsmouth		48-0	17-6	7-7	74	Sold 14 Apr 1698
Charles	8	30	1675	Rotherhithe		54-0	20-0	7-9	120	Cast away on the coast of Holland, Nov 1678

Ships built for the Royal Navy 1660-75 and their builders.

RATES	1st	2nd	3rd	4th	5th	6th	Unrated	Total
Deane	3		4		3	6	9	25
Phineas Pett	2		2	2			13	19
Jonas Shish	2		2				12	16
Christopher Pett		2		1	1		3	7
Sir John Tippets		2					3	5
Other Builders	1	1	2	5	3		10	22
Total	8	5	10	8	7	6	50	94

Source: Pepys' 'Register of Ships', in Navy Records Society, Vol 26, Pages 266-299.

NOTE
This is not intended to be a full biography of Deane. For further information about his life, see the article by A W Johns in *The Mariner's Mirror*, Vol XI, pages 164 - 93. The entry in the *Dictionary of National Biography* is brief, and rather unbalanced.

Much of our information about Deane comes from Samuel Pepys, and is to be found in his *Diary*, in the volumes produced by the Navy Records Society (*Calendar of Pepysian Manuscripts*, Vols 26, 27, 36, and 57, *Naval Minutes*, Vol 60, and *Tangier Journal*, Vol 73), and also in collections of his letters, especially *Further Correspondence of Samuel Pepys*, edited by J R Tanner, 1929, and *Shorthand letters of Samuel Pepys*, edited by E Chapell, 1933.

These largely give Pepys's personal view of Deane, although they also give a great deal of information on his work. This needs to be counterbalanced by a more objective assessment, which is to be found in the *Calendar of State Papers, Domestic* especially between 1660 and 1673, which also include Navy Board papers, and give a great deal of information on shipbuilding policy of the time, sometimes augmented by reference to the original manuscripts in the Public Record Office.

The best discussions of early ship design are to be found in *The Search for Speed under Sail* (H I Chapelle, 1967), and in 'Early Seventeenth Century Ship Design' (William A Baker. *The American Neptune*, Vol 14, No 4, pages 262 - 77).

Early works on ship design include *A Treatise on Shipbuilding, c1620*, edited by William Salisbury (Society of Nautical Research Occasional Publications, No 6), and Matthew Baker's *Fragments of Ancient English Shipwrightry*, which, like Deane's *Doctrine*, is in the Pepysian Library.

Sir Westcott Abell has published some extracts from both Deane's and Baker's works (*The Shipwright's Trade*, 1948; reprinted 1981), but I disagree with much of his interpretation of Deane.

Top right: A 70-gun ship, probably one of those of the 1677 programme. The forward rake is much shorter than in Deane's *Doctrine*. *(Rigsarkivet, Copenhagen)*
Bottom right: An early ship's draught, probably of an 18-gun 6th Rate of around 1670. An early example of the use of water-lines. *(National Maritime Museum)*

INTRODUCTION

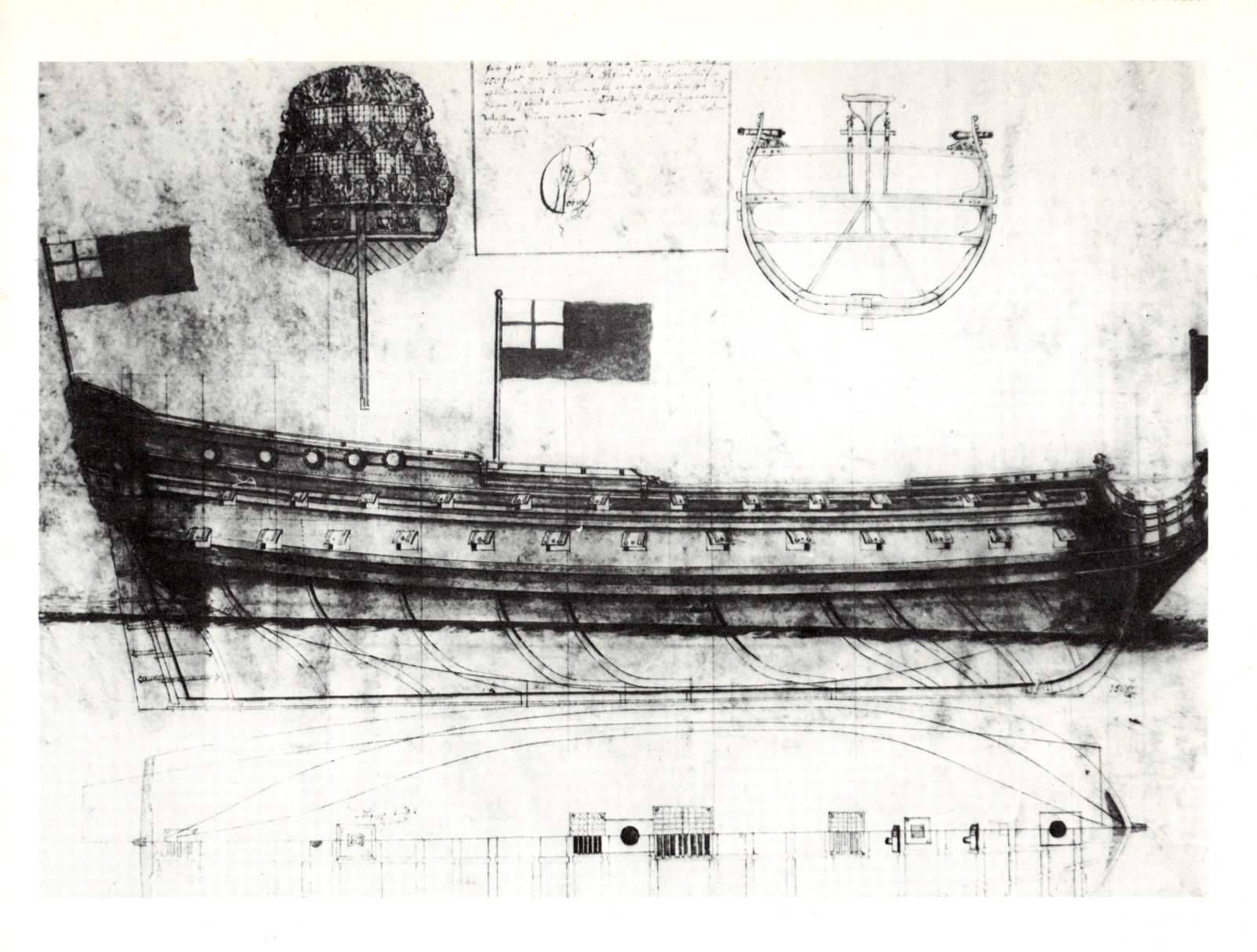

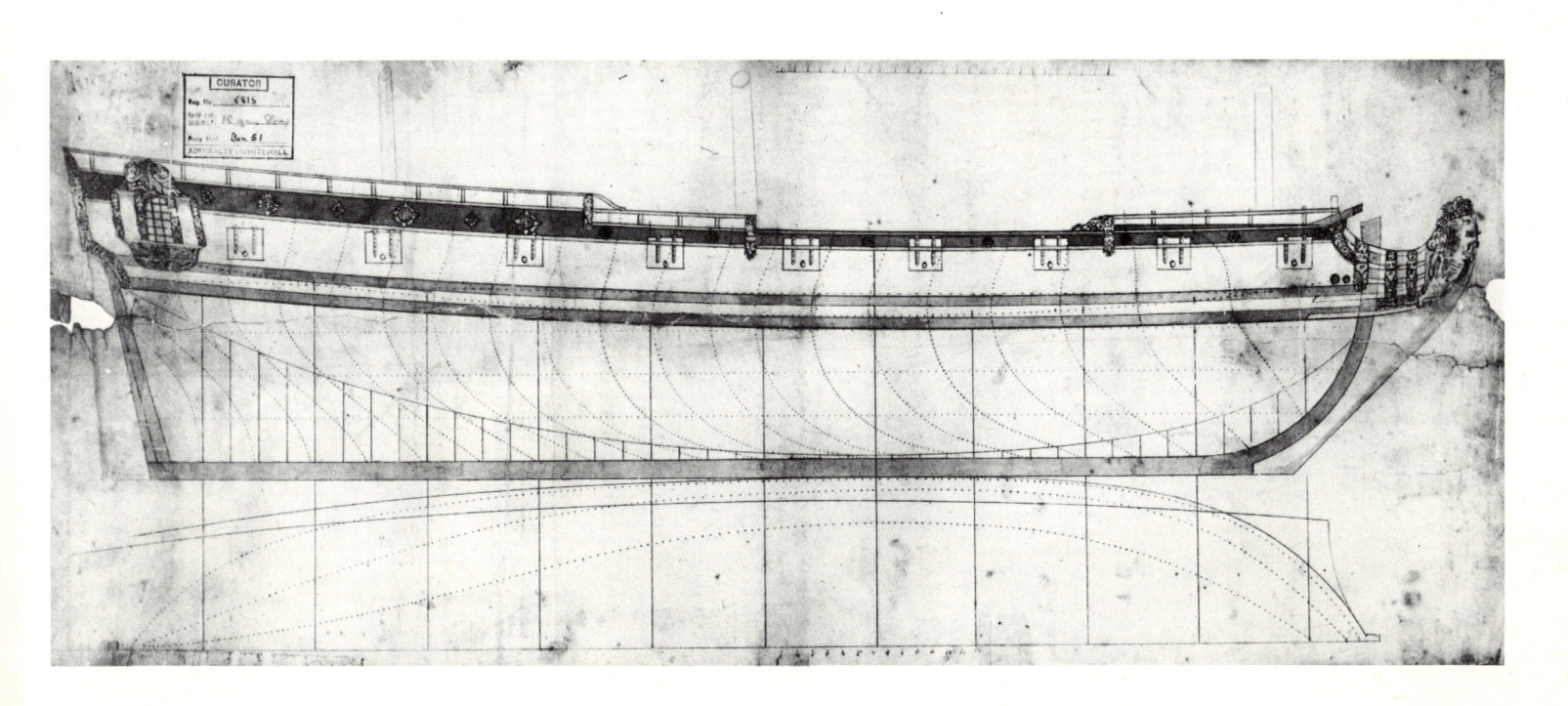

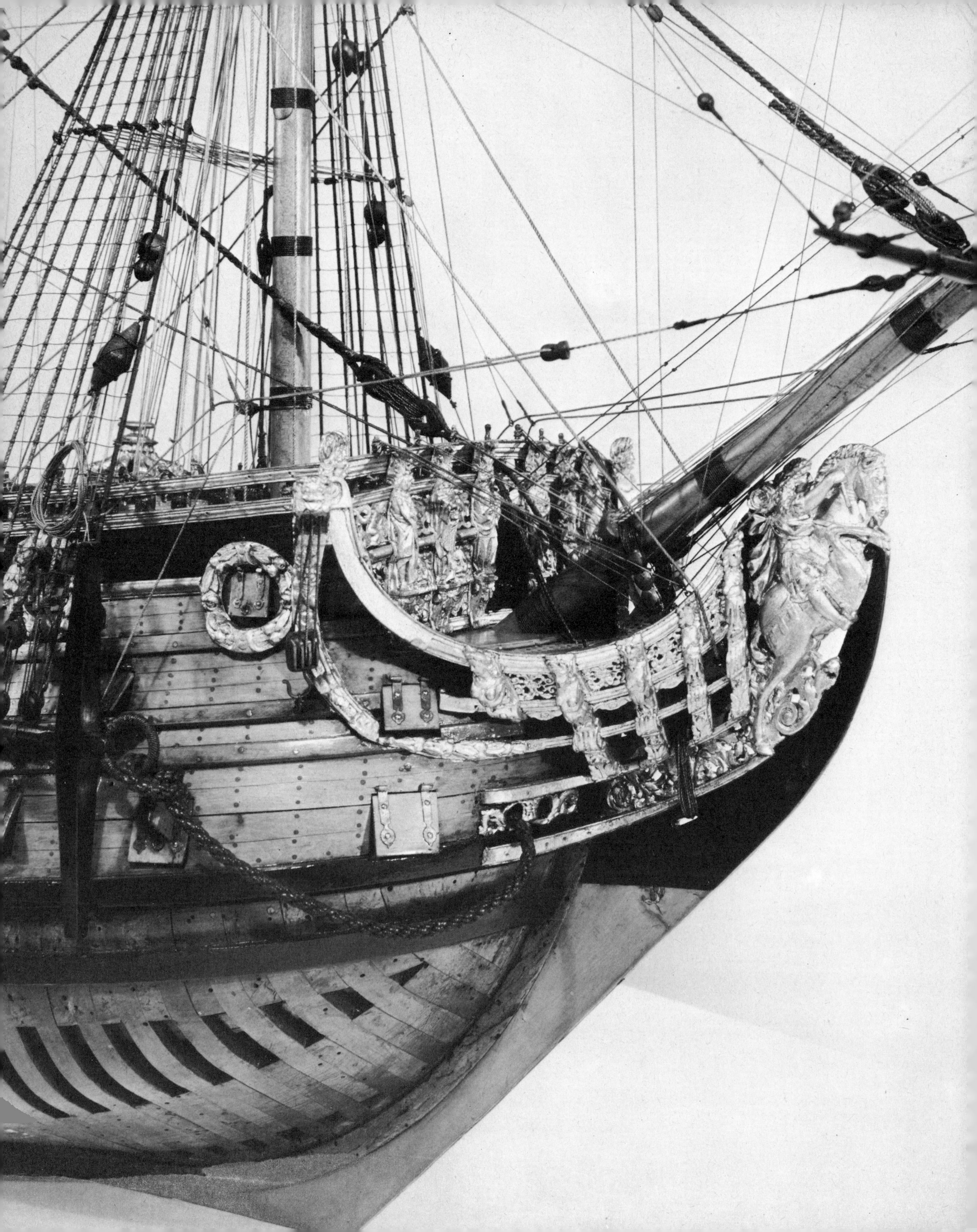

DEANE'S DOCTRINE OF NAVAL ARCHITECTURE 1670

Part 1 Arithmetic

NOTES ON THE TRANSCRIPTION
The original of Deane's *Doctrine* is in the Pepys Library, Magdalene College, Cambridge, manuscript No 2910. Most of the work for the present transcription was done from the photocopy in the Manuscripts Department of the National Maritime Museum, Greenwich, with occasional references to the original to check doubtful points.

I have modernised the spelling, and added punctuation in order to make the text more intelligible to modern readers. In many cases Deane gives details of his calculations in the text. These have mostly been omitted, for they are unnecessary in the age of the electronic calculator. Fractions are also put in a modern form; for example where Deane writes '9 feet and $\frac{648}{1728}$ parts of a foot' I have substituted '$9\frac{648}{1728}$ feet'. Contractions, such as 'pd' for 'perpendicular', have in most cases been expanded. Inconsistencies in the tables have been regularised, but although dashes (—) have been inserted where there are blanks in the originals, no information had been added.

The order of the pages has been altered, in that the section on rigging comes at the beginning of the original manuscript, but it seems more logical that it should be placed at the end. This may well have been Deane's original intention, for the book was clearly rebound after it came into Pepys' possession, as the title page refers to 'Sir Anthony Deane', though Deane was not knighted until 1675. The original page numbers are given in the margin and all references in the notes quote these numbers. Any apparently 'missing' pages contained the drawings. The page headings are taken from the contents page of the original, which has not been reproduced in itself. The division into three sections, 'Arithmetic', 'Hull Design', and 'Rigging', is mine, not Deane's, and there is no trace of it in the original. Apart from the exceptions mentioned, the text is complete, and unaltered.

PART 1. ARITHMETIC
It is possible that this section was added after the section on hull design, since some of the information is duplicated in the latter section, For example, the method of drawing a perpendicular is given on page 22, and again on page 39 of the original. The methods seem elementary, and sometimes crude, by modern standards, but it should be remembered that many shipwrights were almost innumerate, including even Master Shipwrights in the Royal Dockyards, as Pepys often complained. The calculations given are mostly highly relevant to a shipwright's work, either in the drawing of the draught or in the measuring of timber, which was an important part of his responsibilities. Some of them could be used to calculate the weight of the hull, but Deane does not provide a complete method for doing this to match his method for calculating the displacement. He does not tell us how to find the weight of a complex shape, such as one of the timbers making up the frame.

Left: The head of the *Prince* of 1670.
(Science Museum)

The demonstrating of a prick, a straight line, a circular line, and to raise or let fall a perpendicular, the one line parallel to another.

p21 Before I proceed unto the several lines in a ship's body, I do think it not amiss, for the sake of those who may be desirous, to learn what is intended for instruction, namely to show them the full art of measuring all sorts of figures and shapes, both superficial and solid, together with the art of gauging, and the rule static, as well by geometry as by arithmetic, which being attained unto, I conceive those who intend to build ships may be the more apt to proceed without difficulty. So that it may be better understood, I do begin in a short method, only laying down one demonstration in every part of my work, until I have finished what I have promised, whereby those who have not the use of arithmetic may be gainers by their labours, as those who have made great use thereof. And so proceed unto the laying down to a point, which is the first work to be understood in our art. A point or prick is the least dimension and cannot be divided, it being the smallest matter you can conceive in your mind or demonstrate, and its use is mostly, in this work, a centre whereby you describe either circle or parts, and is demonstrated as you find it, by a small prick with the letter A.

Having showed you a point as aforesaid, do proceed unto a line. A line is length and not breadth or thickness, and may be divided from A to B equal or unequal, as from A to C unequal, and from C to B equal, as for example ACB. From A to C the line is divided into three unequal parts, but from C to B into two equal parts, and thus you are shown what a line is, namely a straight line, which is the shortest of lines. There are other lines, as circular lines, whereof great use will be made in our work, and therefore I shall show what is called a circular line, as, for example, take a pair of compasses and set one leg at A and open them at any distance, and with the other leg of your compasses describe an arc or a piece of a circle, and that line is a circular line, or a segment of a circle, as the line BCD. Which is all I shall say as to lines straight or circular, conceiving these appear very plain. I do proceed to the next, having already showed what a straight line is and also a circular, as is manifest by the two problems in the margin.

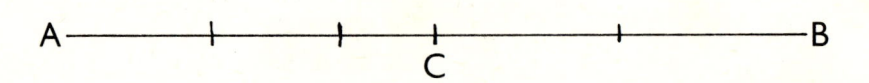

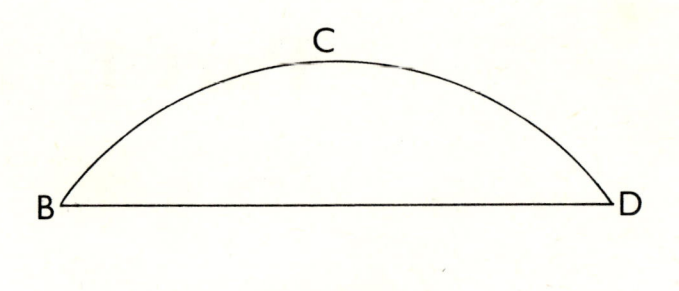

A circle is performed by opening your compasses at any distance. Setting one leg in A, strike a line beginning at B. Continue it round A until it meet at B.

p22 To Raise or let fall a perpendicular upon any right line given. Let AB be a line given and let C be a point therein, whereon I would raise a perpendicular. Open your compasses to any convenient distance and, setting one foot in the point C, with the other leg mark on either side thereof the equal distances as

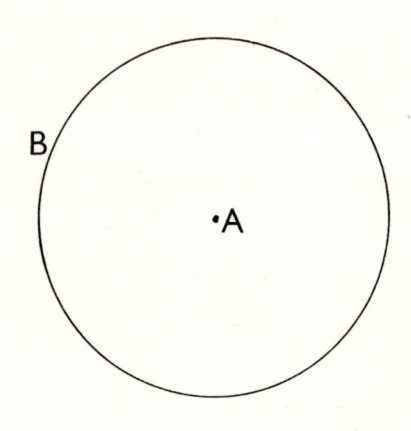

CE and CF. Then, opening your compasses to a convenient distance wider than the former, setting one foot in the points E and F, and strike two arc lines crossing each other as in D, from whence draw the line DC, which is a perpendicular to AB, or as we call it, a square line to the line AB.

To let fall a perpendicular from a point assigned to a line given, as for example let the point given be D, from which I would let fall a perpendicular, and let the line whereon it should fall be AB. Open your compasses to any convenient distance, and, setting one foot in the point D, make an arc or piece of a circle with the other foot until it cut the line AB twice, that is at E and F, then find middle between these two intersections, and from that middle draw a line to C from the point D, which is the point given, and that line shall be perpendicular or plumb from the point D unto the line AB as was required to be done, and will be often in use for the work we are to proceed in building, and therefore requires your memory of this easy but useful demonstration.

To a line given to draw a parallel line at any distance required. Supposing you now well understand a line straight or circular, and also a perpendicular, you shall be instructed in the drawing of any lines parallel one to the other, as for example I would strike a line parallel to the line AB. Open your compasses to any convenient distance. Set one leg at A, and with the other strike an arc at C. Again set your compasses, keeping the same distance, at B. Strike an arc as before. Having so done, strike a line at the extreme of both arcs and it will be the line C and D, parallel to the line AB, both lines being of equal distance from one another in every part, which was required to be done. In understanding this method, all parallels are of like nature, let them be perpendiculars or bases. The work is all wrought as by this, above said.

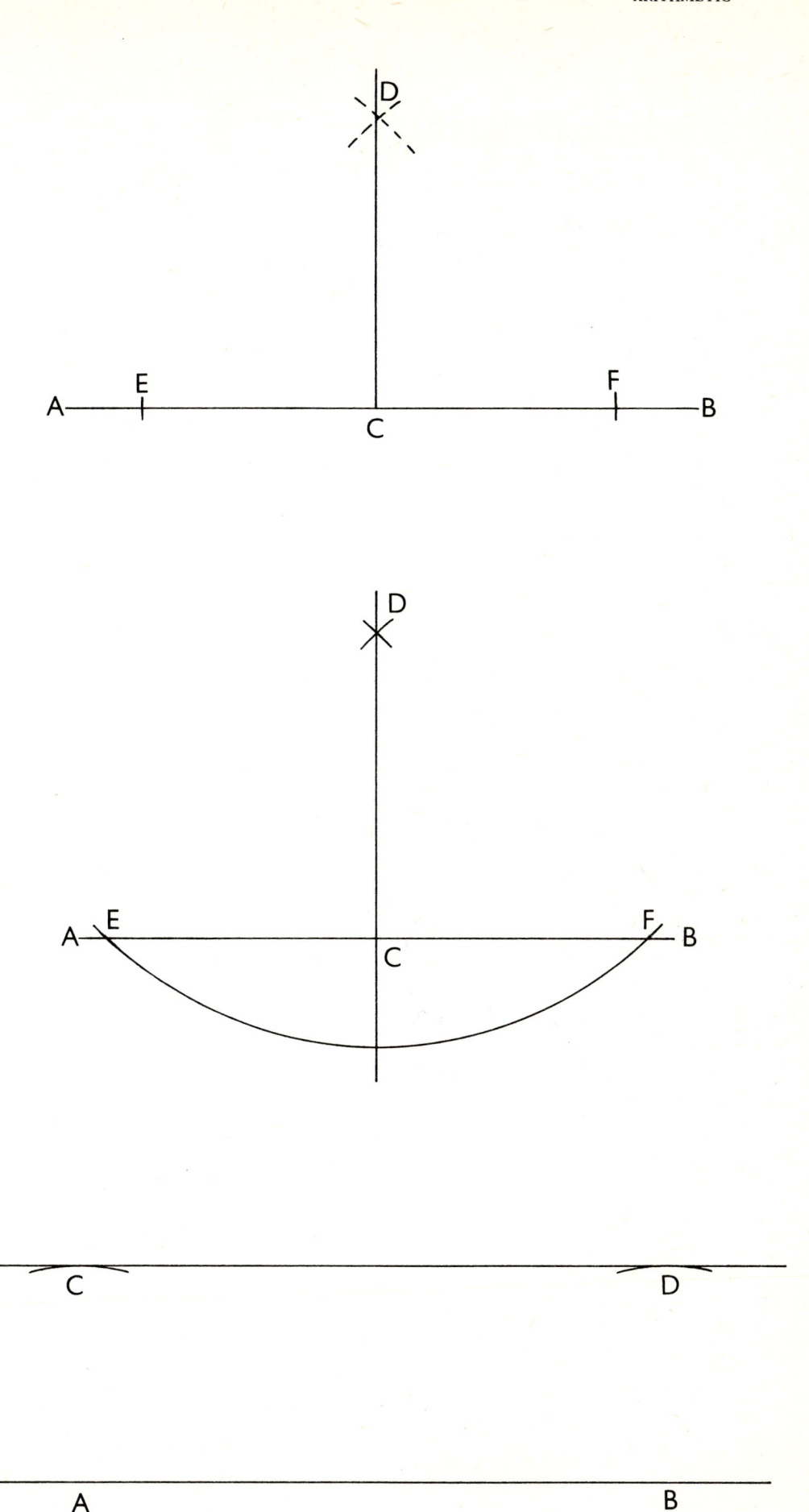

The demonstrating of superfices, and also three several angles, the rectangle, obtuse and acute.

p23 To describe a superfice[1]. Having as I conceive made you perfect in the knowledge of lines, it remains to instruct you in the uses thereof, which being applied makes a superficies of several sorts, as for example strike a straight line AB, whereon raise a perpendicular at A to C. Open your compasses at such distance as you please, setting one leg at B. Strike an arc parallel to the line AB, which is E. Having so done, keep the same distance, setting one leg at A. Strike an arc from A to C, then strike the line CE, which makes the depth of this figure, and for the length as you please, doing as you have been taught. By another parallel make the line BE from the line AC, and thus you have a figure of four sides, which being only length and breadth without depth is called a superficies, or plane, contained under four sides. There are many various shapes as the lines are applied, and therefore I shall show you others, for example suppose a straight line AB whereon is raised a perpendicular line AC. Strike a third line from C to B, and the figure so prescribed is a three sided figure or superficies, and is called a right angle[2] for shape, which having added unto the same quantity makes a square. And then is the line CB a diagonal line, and thus may you treat as many shapes as may serve your turn until your work ends.

Before I proceed to tell you what an angle is, I shall give you a demonstration of a square figure which is four equal sides, the one side not exceeding the other, and is of all things necessary to be known. The way of raising is by drawing a line as AB and raising a perpendicular on each end at A and B, of length equal to the line AB, and as you have been taught set one leg in A and strike a small arc, doing the same at B, then strike the line EF, and thus have you a figure of four equal sides, which is a due square in every way. In which figure, if you strike a line from corner to corner, it contains two right angles, if from the four corners the lines crossing each other there will be contained four angles, so that this figure is called a quadrangle as well as a square, which was required to be shown, as is more plain in the figure ABCD.

This square figure divides a circle into four equal parts. As for example, draw a circle, wherein draw a line crossing the centre and it will divide the circle into two equal parts, as you see the line AB divides the circle equal. Then raise a perpendicular on the centre and strike a line from E to F and it will divide the circle into four equal parts. Each part is commonly called a quadrant, which was required to be known.

p24 To describe an angle. There be several sorts of angles, but none so useful for our work in hand as those which shall be laid down, namely rectangles, obtuse angles, and acute. Nor is there anything more beneficial than the ground of these unto the young artist. Therefore I will set them down in their several shapes from which they receive their names, beginning with the right angle. Suppose a line for your base is AB. Raise a perpendicular from C to D. Having so done, your angle is made by the perpendicular or plumb line at the point C, and hath his name of right or rect- angle, from its upright line which divides a circle into four equal parts, for whatsoever is perpendicular unto any line divides each side of the line alike, nor is an angle any more than two lines meeting together, and is distinguished by dividing the circle equal or unequal as shall be shown. This is therefore a right angle. The next shall be an obtuse.

---

1. *Superfice.* Surface.
2. *Angle.* This word is used in two senses, in the modern sense and to mean a triangle.

The obtuse angle is greater than the right angle by so much as it extends itself from the perpendicular line you last made in your rectangle. As for example, draw a line as AB, raising a perpendicular at C to D. Having so done, open your compasses at any distance not exceeding the length of your lines from D to E, then draw a line from E to C, and then have you an obtuse angle ACE, forasmuch as the angle ACE is wider than the rectangle ACD, and contains the angle DCE more than the angle ACD. Which was required to be done, and thus you see that the obtuse angle is all ways greater than a rectangle. It remains to show an acute angle as in the next shall be described.

The description of an acute angle. An acute angle is less than the rect- or obtuse angle. As for example, draw a line as AB, whereon raise a perpendicular CD. Having so done, open your compasses as in your last work, setting one leg in D to E, then draw the line from E to C. It makes an acute angle BCE, and is less than the rectangle ACD by the angle DCE, so that by this you may plainly perceive the nature of the three angles apart. The obtuse biggest, the acute the least, and the rectangle the mean. So that there needs no more concerning the knowledge of these, only, by the way, it is not amiss to show you that the several angles in one figure, as you see here demonstrated by the lines, ACD is a right angle, as you have been taught, the lines ACB an obtuse angle, the greatest, and the lines BCE an acute angle or least, and thus have you the definition of these three angles, and the way of prescribing them by geometry, which you must know ere you pass unto our works, which will be perceived in its due order. Hoping you well understand now the three several angles, do pass unto the making of a scale.

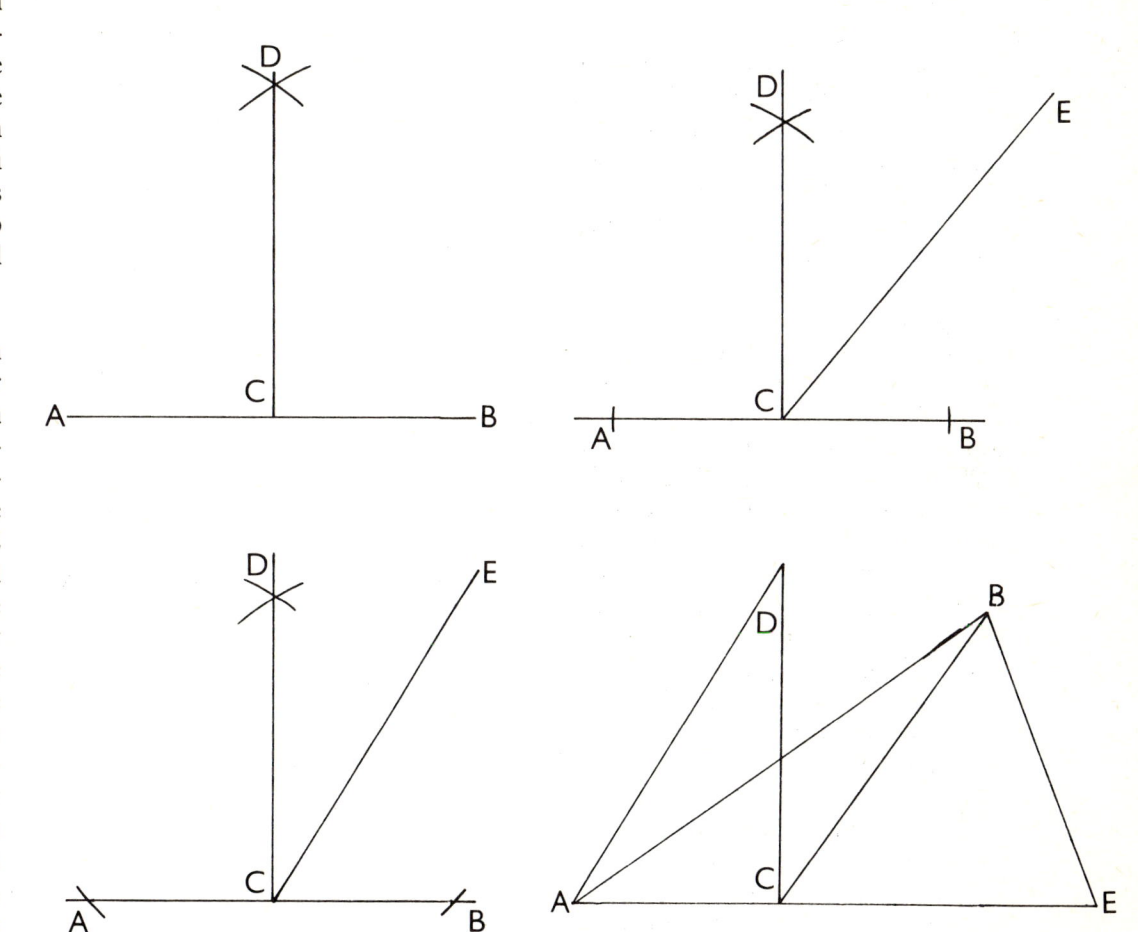

The demonstration of a plain scale, by which you may work several lines in proportion one to the other.

p25 It is necessary before we proceed any further to make a scale to enable us the better to lay down our geometrical problems. Besides, it will perfect the young learner in its use for shipping, without which he cannot well proceed. And to make it more easy perceived, it is a piece of board, brass or paper, whereon you make with your compasses several equal parts, greater or lesser as your plate or design will afford, the larger the better regarding the design in hand, and considering a ship, a body of many feet, which will take up a great room. I choose to suppose a quarter of an inch to a foot, or as my plate will admit of, more or less. This being done, I strike two straight lines of equal length, which I divide into 24 equal parts, beginning at 1,2,3,4,5,6, until I have procured 24. When it is thus divided, subdivide every part into eight parts to make it into half quarters of each part if your scale will admit thereof. Having now made your scale, it is applicable to either miles, roods, yards, feet, or money, as pounds, shillings, pence, labour, and the like. Therefore it is of absolute necessity to be very exact in the making of it, by reason of error which will other ways attend your work. Being thus prepared by any scale of equal parts, I shall proceed in the next place unto its use, this being only to prescribe the scale and its shape, which was required to be demonstrated.

Note that a scale is only equal parts, and it in proportion to whatever you apply it. If you apply it to inches it bears proportion as 1 to 4. This scale being $\frac{1}{4}$ part of an inch, if to feet it bears as 1 to 48. If to yards, it is as 1 to 144, and so accordingly to whatever you apply it, as you shall find hereafter.

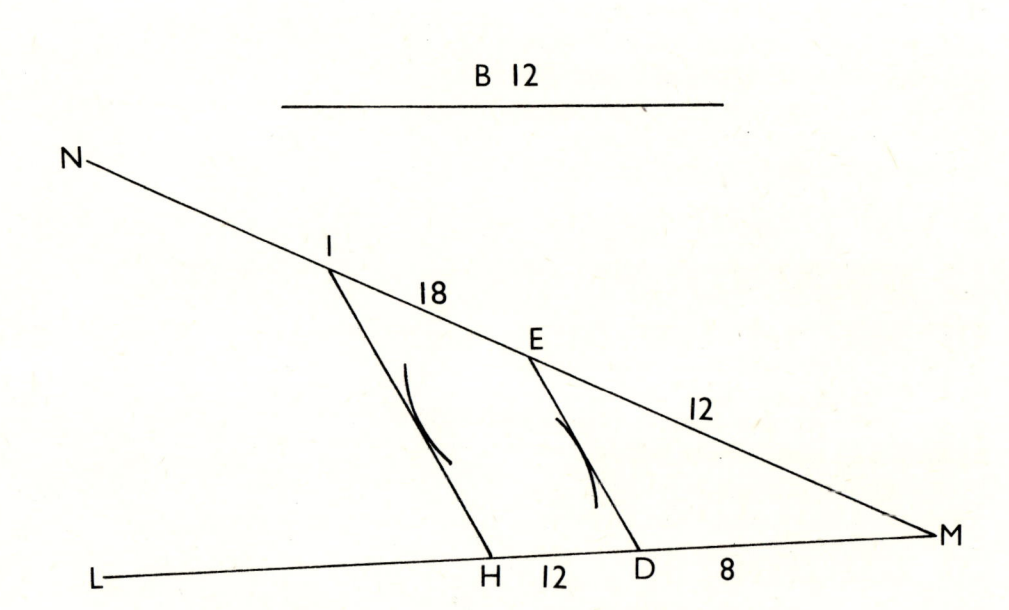

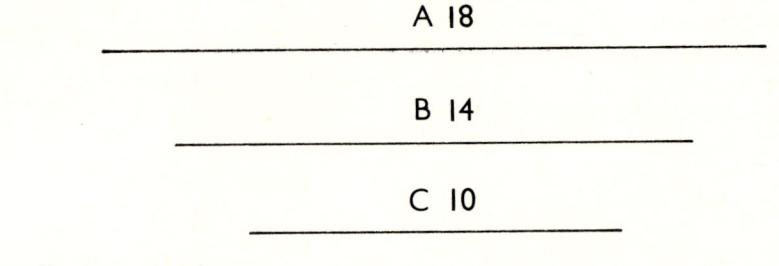

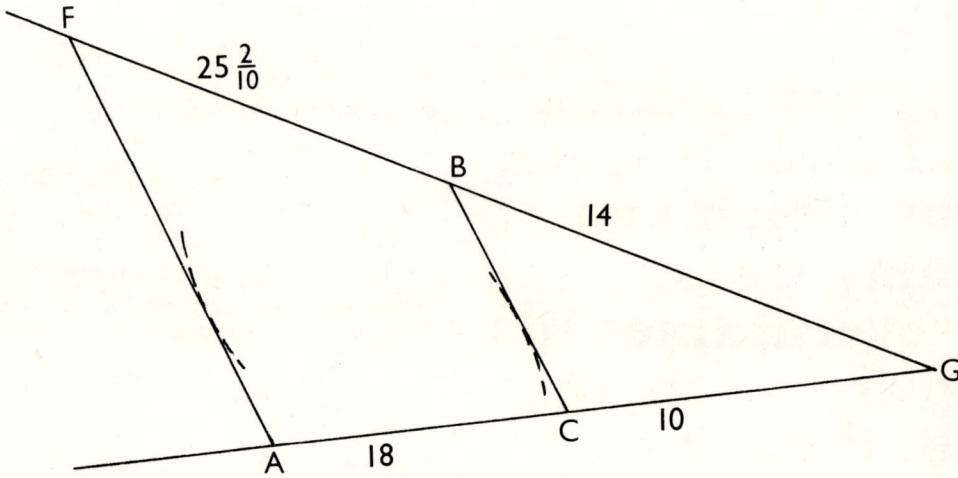

p26 Having two lines given, to find a third proportional line to them. Your scale now having been made as on the other side, I shall proceed to the drawing of lines in proportion the one to the other, as for example, the line A is eight feet, and the line B is 12 feet. I desire a third line in proportion to A, as A is to B. You must work thus. Strike at adventure[1] two lines meeting at one end as an angle. Let the two lines be LMN. Take from your scale the line A, which is eight feet, and set one leg at M, the other at D. Having so done, take the line B, which is 12 feet, and place it from M to E, and then draw the line D and E. Having so done, take again the line B which was 12 feet, and set it from M to H. Lastly take the distance from H to D and strike a line parallel to ED, which will be the line HI, 18 feet, and is in proportion from I to M as the line B is to A, as appears by the figures demonstrated in the margin. By this method are all lines drawn by geometry, so that it is needful for the young artist to bear this work in rememberance, and so I shall say no more of this nature, but pass to the rest of the work, in which I will be very brief, only according to my promise.

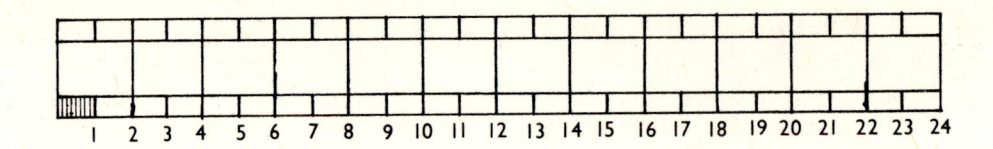

1. *At adventure.* At random.

38

Having three lines given, to find a fourth proportional line to them. You must proceed in this work as in the former, striking two lines at adventure, like to an angle meeting at each end, supposing them to be the lines KGD, meeting at G.[2] Having so done, the question which is proposed is to have three lines given to find a fourth, as the lines ABC, the line A being 18 feet, the line B 14, the line C 10. I would have a fourth line in proportion to A as B is to C. Work thus. Take with your compasses the line C which is 10, and set one leg of your compasses at G, setting the other leg at C. Having so done, take the line B, which is 14 feet, setting one leg at G, the other at B, and then strike the line BC. Having so done, take the third line A, which is 18, and set one leg at G and the other at A. Then strike a parallel line to BC, which is the line AF, and then is the line from F to G $25\frac{2}{10}$, which is the fourth line sought for, being in proportion to A as B is to C. And in like manner is all work of this nature concluded, which I pray remember before you pass into the next.

By a plain scale to measure board or timber of equal or unequal sides without the use of numbers

p27 How to measure any board geometrically. In the last you were instructed how to find any line in proportion, which is like unto this, only in your last you had three lines given, to find a fourth, but in this you have three numbers whereby the fourth is found. Only you must observe that the first of the three numbers is always twelve, being the side of a square foot of board, or the side of a cubical foot of timber. The second number is always the number of feet contained in the length of the board. The third number is always the number of inches contained in the breadth of the board, and the fourth the square feet contained in the whole board, which is the resolution of the question. As for example, I would know how many feet of board is contained in a board 18 inches broad and 14 feet long. Work thus. Draw two lines at any distance, meeting at one end as you have been taught, as suppose the two lines ABC, which meet at B. Then take 12, your square inches in a side of a foot, and place it from B to D. Having so done, take the length of your board, which is 14 feet, and place it from B to E, then draw the line DE. Having done that, take the breadth of the board, which is 18 inches, placing it from B to F, then take the distance from D to F and draw a line parallel to DE, and it will be the line FG, which from G to B is 21 feet, the content of the whole board. All board is measured in like manner, which is worthy of your observation.

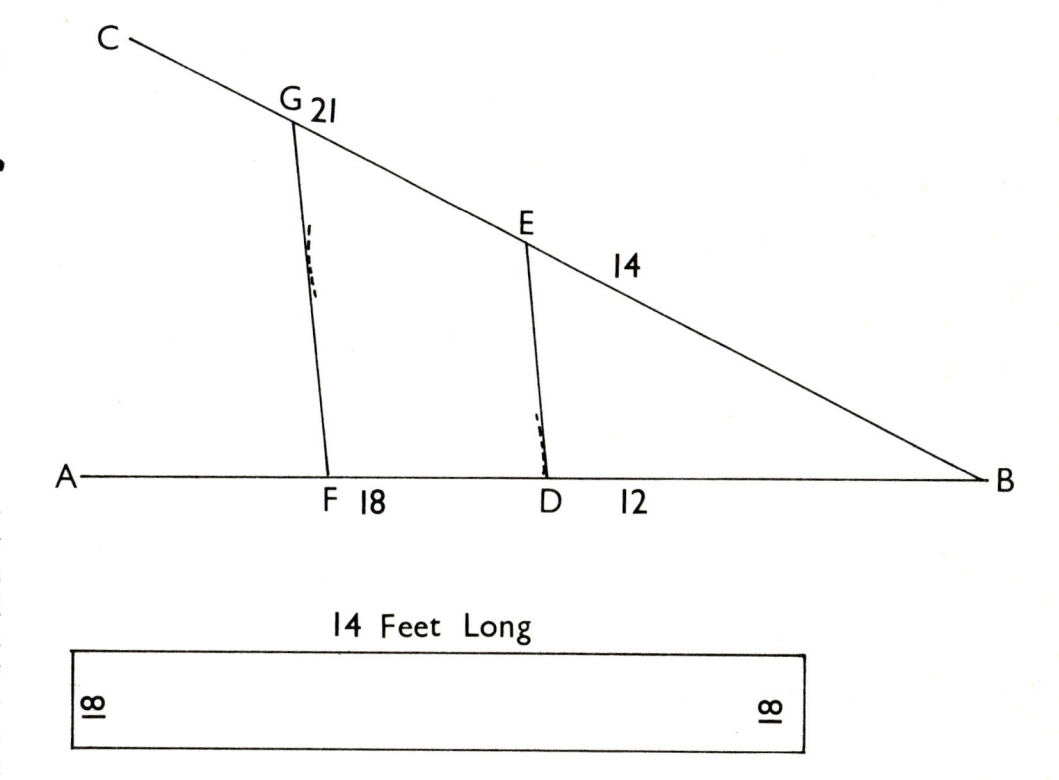

To measure any board under a foot broad. In the last you were shown how to measure any board, stone, or glass which was above a foot in breadth, but in this you shall be instructed how to measure any board, stone, or glass under a foot in breadth, by reason the work differs from the other. As for example, suppose a board of 16 feet long, and 9 inches in breadth. I would know the content thereof in feet. Work thus. Draw two lines as before you were taught, which let be ABC. Then take from your scale of equal parts 12, which is your side of a foot square, and set one

---

2. *G.* This is left blank in the original, but it is clear from the diagram that this is the letter meant.

leg of your compasses at C, the other at D. Then take your length, which is 16, and set it from C to E. Then strike the line D and E. Having so done, take your breadth, which is 9 inches, and set it from C to F. Then take your distance from F to D, and draw a line parallel to DE, which will be the line GF, and then is the line GC 12 feet, which is the resolution of the question, and was required to be done. For in your last question the fourth number exceeded your first work, but in this it is within your first, and the reason is because the last work was above 12 inches broad, and this work, under, which is but 9 inches. And thus have you all measures which are flat, remembering always this — as 12 is to the length in feet, so is the breadth in inches to the content in feet.

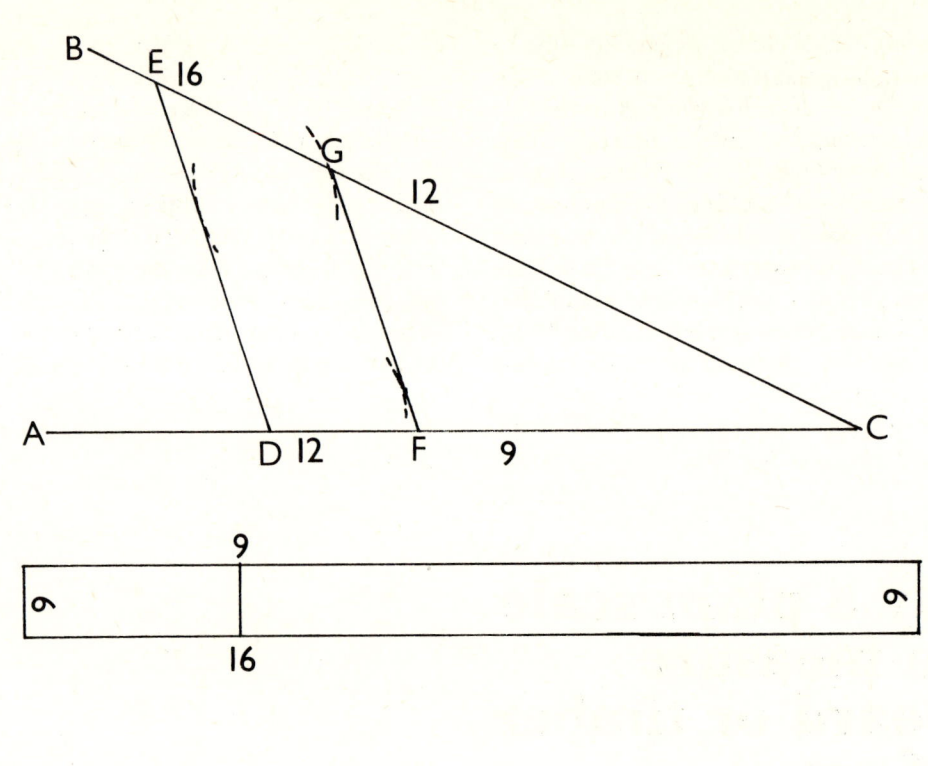

p28 To measure any timber geometrically. The beginning of this work is like unto that of board, and differs in the last line, as you see. For example, I would know how many feet of timber is contained in a piece of timber which is 14 feet long, and 15 inches square. Work thus. Draw two lines as you have been taught, and let them be ABC, meeting like an angle at B. Then take from your scale 12, which is always your side of a foot square, and place it from B to D. Having so done, take 14, the length of the piece, and place it from B to E, then draw the line DE. Then take 15 inches, your breadth, and place it from B to F and draw the line FG parallel to the line DE, and thus have you one side as was directed in the board's measure. It remains therefore to find the remainder, as thus. Draw a diagonal line from G to D for your last work. Having thus done, set off the distance from F to H. By a parallel line draw the line FI, which will be found from I to B 21¾ feet, and so much timber is in the piece demonstrated, which was required to be done. And in like manner are all pieces measured whose sides are equal. It remains to show you how to measure timber whose sides are unequal as in the next plate shall be set down.

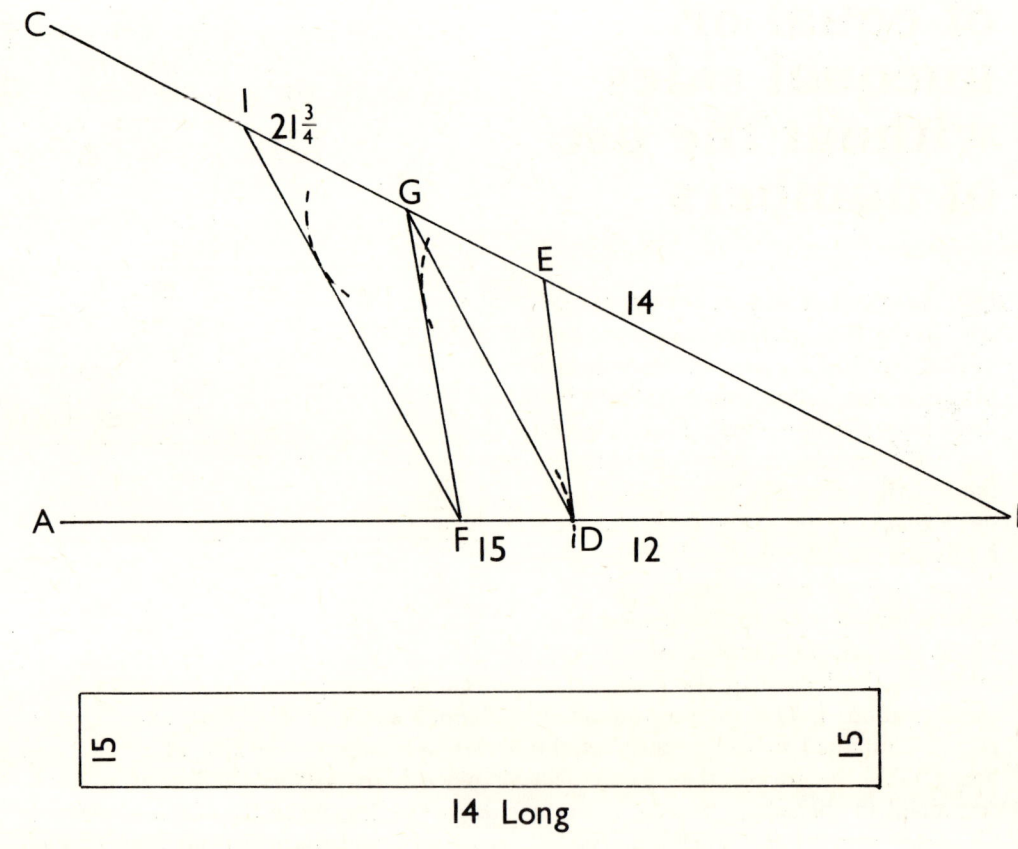

To measure timber of unequal sides by geometry. Suppose a piece whose sides are unequal, the one side being 18 inches, the other side 14 inches, and the length 9 feet. Work as the former chapter, by drawing two lines at any distance, meeting at one end like an angle, and let the lines be ABC, meeting at the point B. Then work thus. Take from your scale 12, which is the side of a foot square, and set it from B to D. Having so done, take the length, which is 9 feet, and set it from B to E. Then draw the line DE. Having done that, take 18, the breadth of one side, setting it from B to

40

E, then draw the line HI parallel to ED. Now have you the superficial content of one side. It remains to find the other, by drawing a diagonal line from I to D, which is the line LM. Having done this line LM, take your side of your scale which is 14 inches, and set it from B to O. Draw another line parallel to ID, which will be the line OP, which distance from P to B is $15\frac{4}{5}$ feet, the cubical feet of timber in a piece of the dimension propounded. And thus have I given you one demonstration both of equal and unequal sides, which was required to be done. I could give many examples more, but considering that these, well understood, are sufficient, I will pass to the next on the other side.

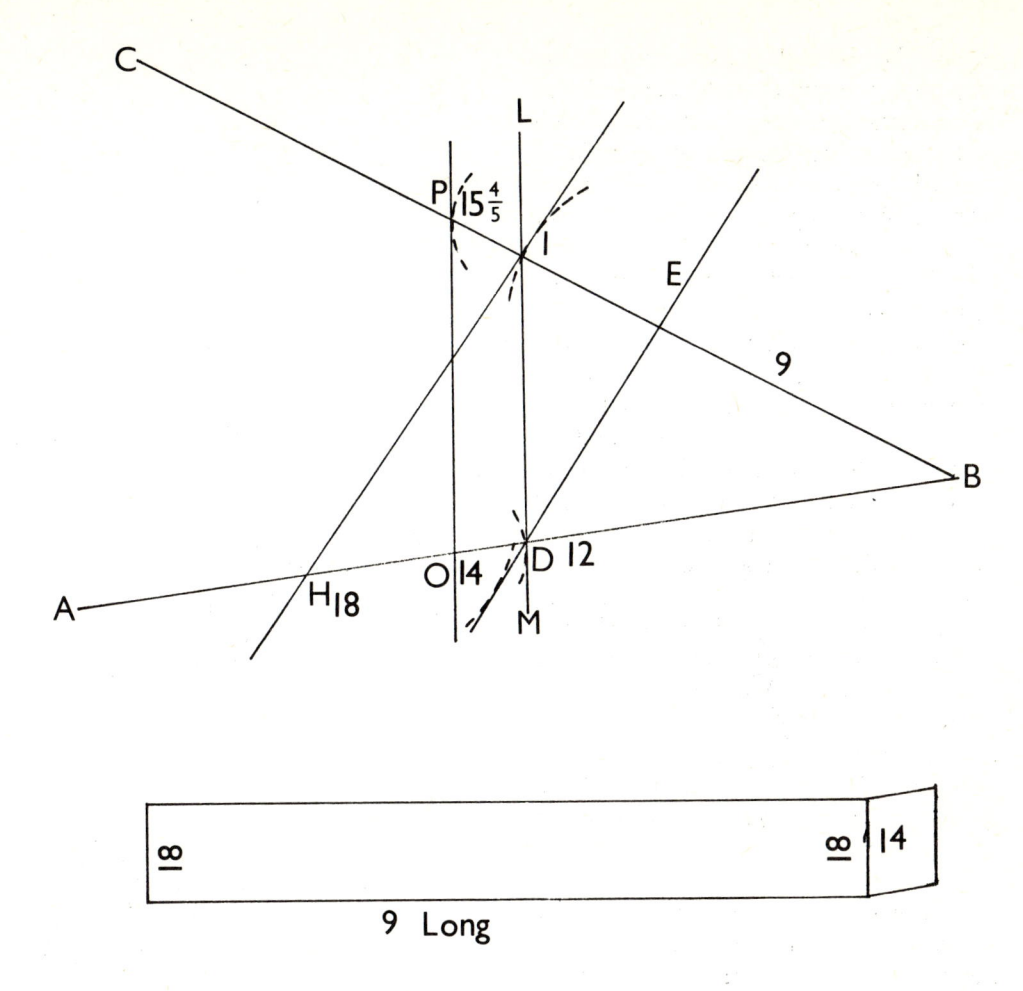

By a plain scale to measure round timber and cast up wages with the use of numbers.

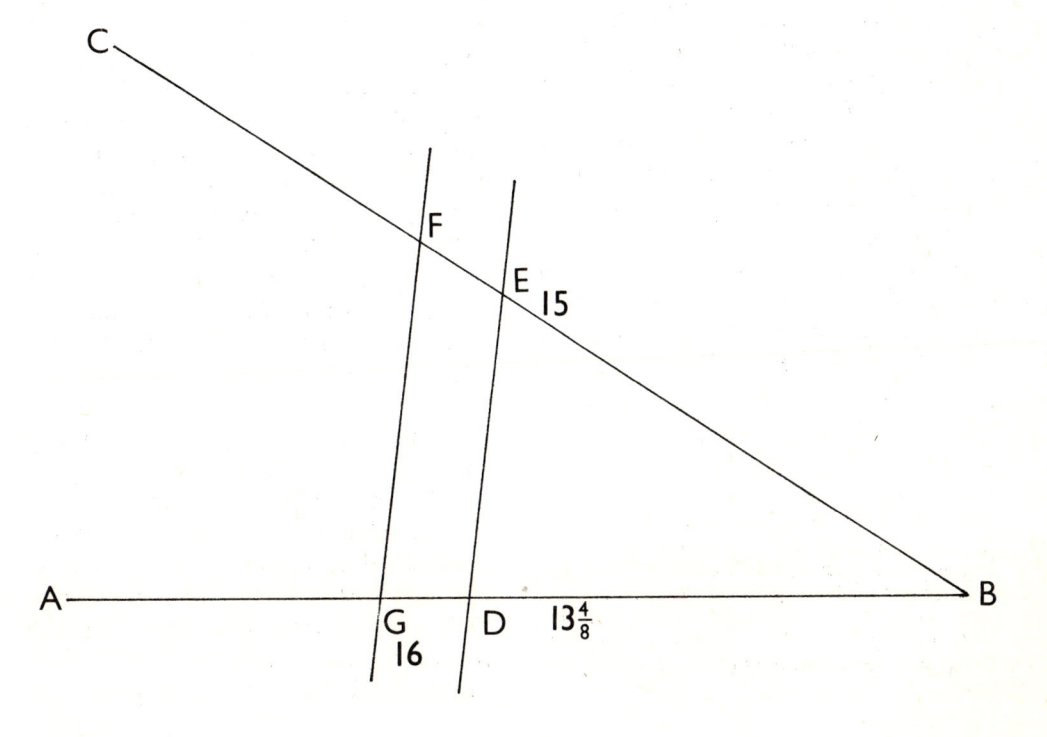

p29 To measure round timber geometrically. In the last chapters you were shown the several ways of measuring board and timber of any dimensions. In this you shall be instructed how to measure round timber, called girt measure, as suppose a piece of round timber whose diameter is 16 inches, and whose length is 15 feet. I desire to know the content of square feet of such a piece, but by the way, pray remember that I here procure you a centre of every round piece, which is 13 inches and near $\frac{4}{8}$ of an inch to work, by reason it would be too much work, and perhaps the young artist not fit to receive it, as shall be shown hereafter. So work thus, as in your former chapter, remembering that the centre of every round piece is 13 inches and very near $\frac{4}{8}$ths, as in square timber your centre is 12 inches. Proceed thus. Strike two lines like an angle, and let them be ABC, meeting at B. Having so done, take from your scale 13 inches and near $\frac{4}{8}$ths, and set one leg at B, extending the other to D. Then take your length, which is 15 feet, and set it from B to E. Then strike the line DE. Having so done, take your breadth, 16 inches, and set it from B to G, and strike a line parallel to the line DE, which will be the line GF, which will prove from F to B $24\frac{1}{2}$ feet, which is the resolution of the question. In like manner is all round timber measured.[1]

To measure any piece by geometry whose end is a triangle or a three sided figure. I would know the timber in this piece, being 16 feet long and 18 inches basis, and 17 in perpendicular. This work differs nothing from the former, only to find the square of the sides, which is always the one half the base and the whole perpendicular, as suppose the line AB, the base 18 inches, and BC the perpendicular, 17 inches. The square is 13 inches, by taking the one half of the line AB, which is 9, and the whole line BC, being 17, added together makes 26, the half of which is 13. Having now found the square of your three sided figure ABC, work as before, striking two lines meeting at A, then take 12, which is always a side of a foot square, and set it from A to B. Having so done, take your length, which is 16 feet, and set it from A to E. Then strike the line EA. Having so done, take your breadth or square, which is 13, and set it from A to G. Then strike the line GF parallel to BE, and thus have you one side. Now, to find the other side of the timber, having thickness as well as breadth, strike a diagonal line from F to B. Having so done, set one leg at G, taking the distance from G until it cuts the line FB. Having so done, strike a line parallel to the line FB, which will be the line HI, whose distance from H to B will be $18\frac{1}{2}$ feet of timber in that piece. And thus much shall suffice. All timber by this application will be truly measured.

p30 To cast up wages by geometry, or any sums of money whatsoever. Suppose that five men earn seven pounds. What shall twelve men earn? Work thus. Strike at adventure two lines, as suppose the lines ABC, meeting like an angle at B. Then take from your scale of equal parts five, and set it from B to D. Having so done, take seven from your scale and set it from B to E. Then strike the line DE. Having thus done, take from your scale twelve, which signifies men, and set it from B to F. Having so done, strike a line parallel to DE, which will be the line FG. And now will it prove from G to B $16\frac{4}{5}$, which is 16 pounds 16 shillings, which is the desired question. In doing this, and observing well its reason, there will be no more need of propounding any more questions, it being very plain that, five men earning 7 pounds, that twelve men will earn $16\frac{4}{5}$, by the rule of proportion. And in like manner is every question wrought, having two numbers to find a third, and having three numbers given to find a fourth, in continual proportion. And therefore, considering my promise, I do pass away from the rules of geometry, excepting the instructing you how to find the centre to any segment or part of a chord whatsoever, which follows my next proposition.

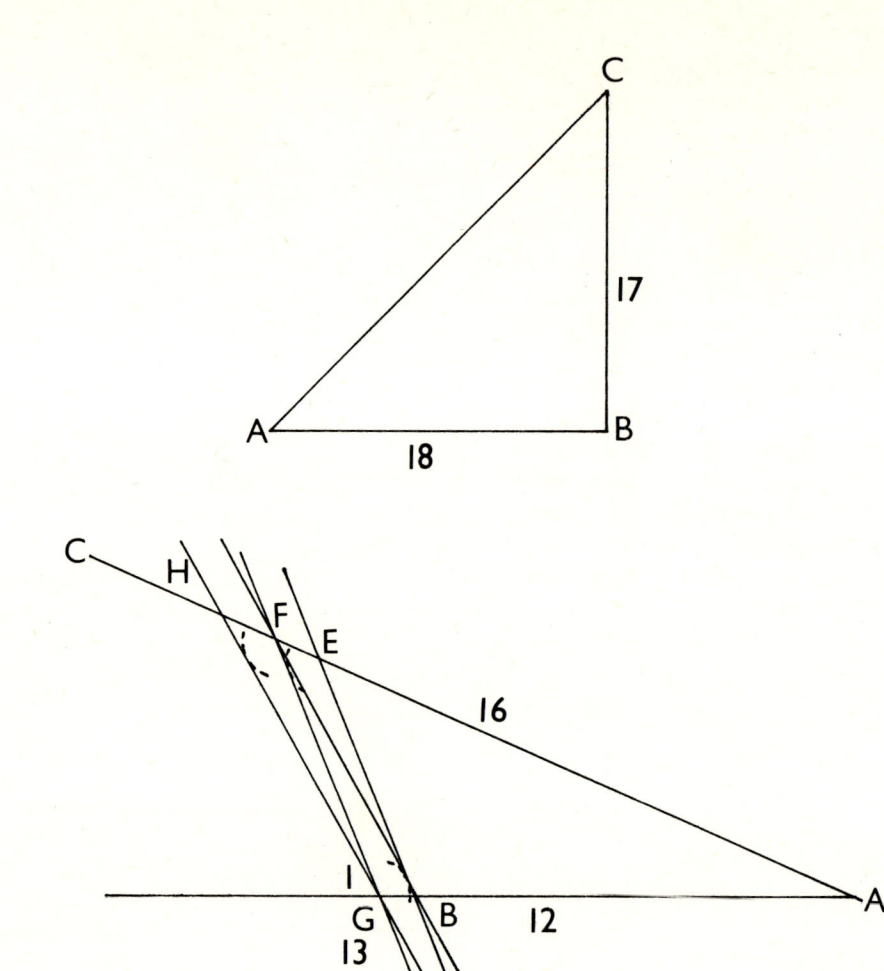

---

1. The whole of this section is confused and obscure. The diagram has several erasures, and it is not clear what is intended. Here it is drawn as described in the text, but in that case it produces a different result from the final figure of 24½ feet given in the text. It is possible that Deane has inadvertently omitted a line from the text.

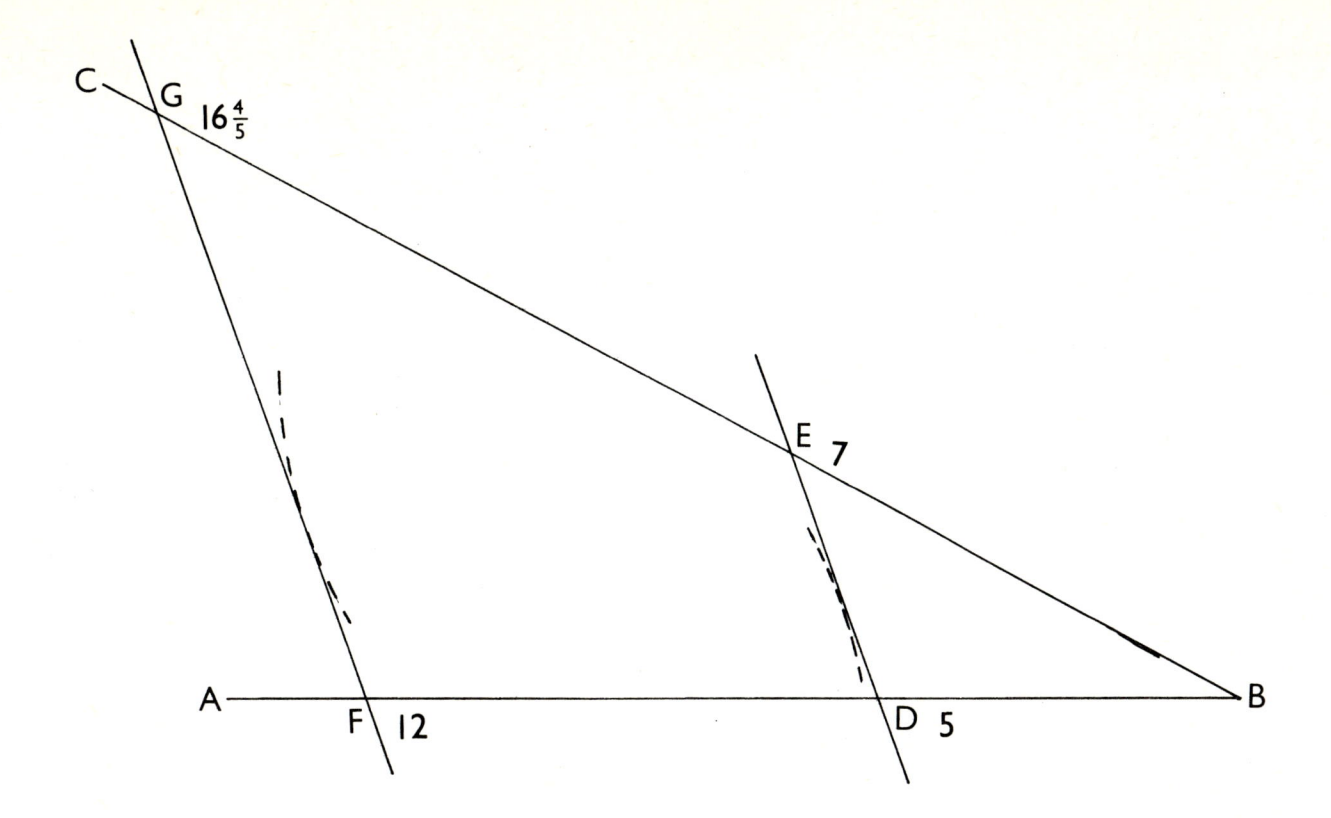

To find the centre of any segment geometrically. Suppose I would find the centre of any segment or piece of a circle, as ABC. First take a point at pleasure with most convenience in the arc ABC, as at B. Now on the point B, at any convenient distance, describe a circle EFGHIK. Which being done, remove your compasses to the point E, where the circle crosseth the arc line given. Now, one foot being set in the point E, and at the same distance as before, cross the circle twice, as at F and K, and with the same distance at the point H cross the said circle twice more, as in G and I, and lastly, by these intersections or crossings draw the lines FK and GI, until they meet or cross each other in the point L, which shall be the centre required. In like manner are all centres to be found out, and therefore shall I trouble you with no more geometrical questions, having laid down all that are useful for our purpose, which I hope the young artist now very well understands. And so shall I pass unto as many questions in arithmetic, that so both may be of use or compared on all occasions, for better manifestation of our whole work, as in the next shall be made plain.

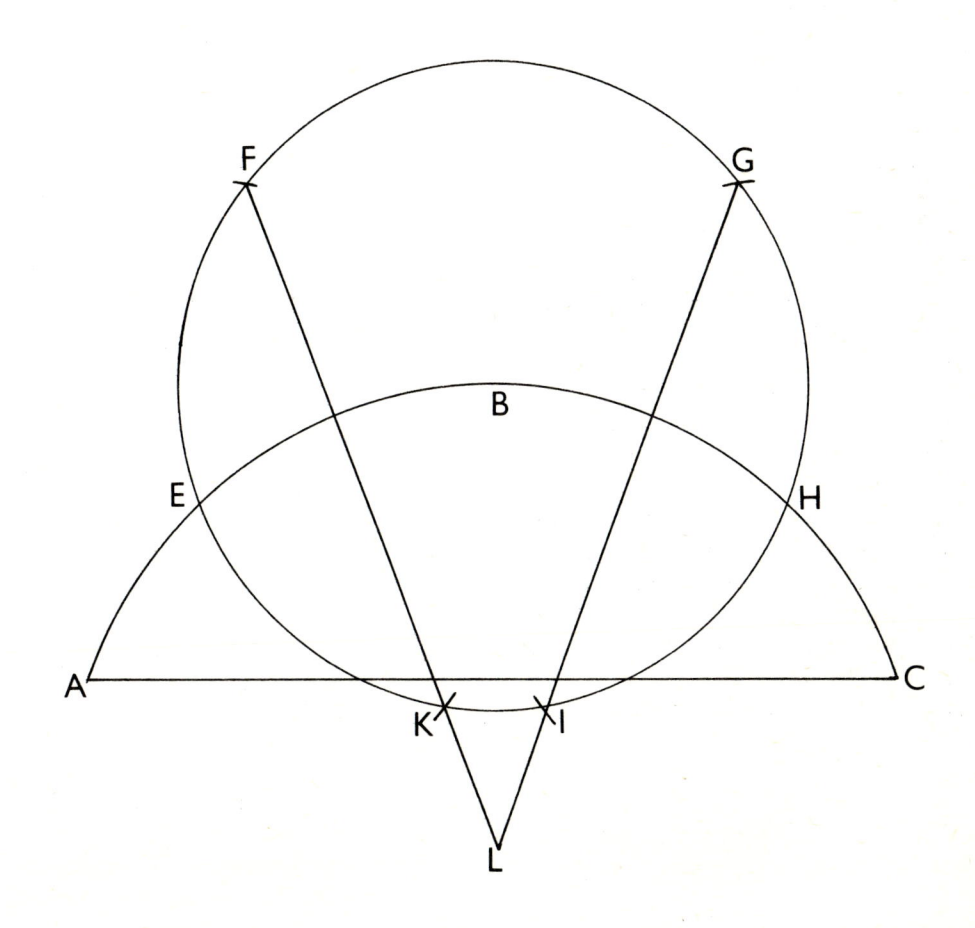

To measure board and timber by arithmetic, being equal or unequal, and to find how many inches in length makes a foot of any breadth.

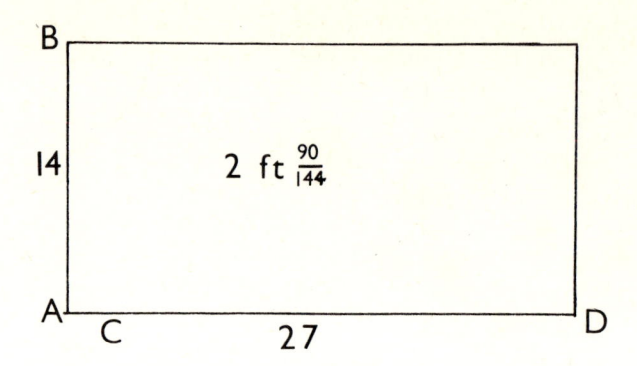

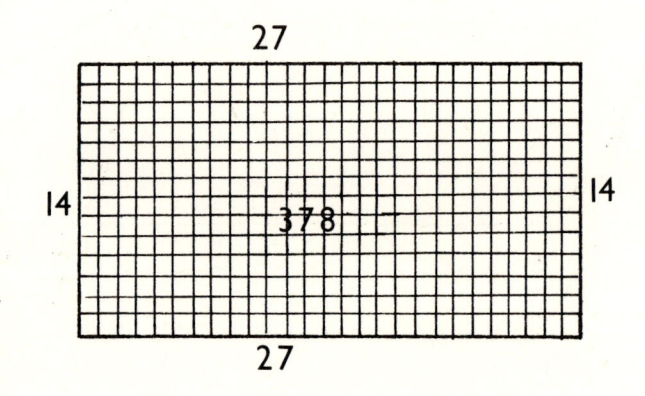

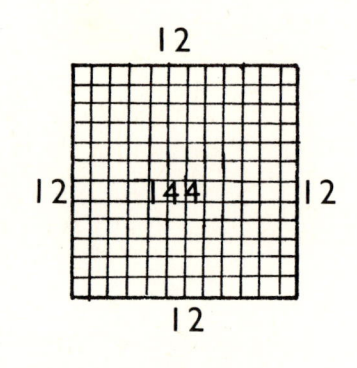

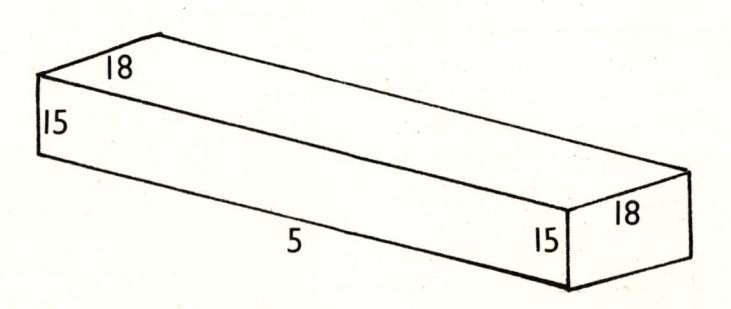

p31 To measure any board, stone, glass, or any measure by arithmetic. I shall be very brief in every part of this work, showing one example to every way of measure, by which all others are performed, and therefore I shall begin with a board or glass, as supposing a board 14 inches broad from A to B and 27 inches long from C to D. I desire to know how many feet square of board is in this figure. Work thus. Multiply 14 inches, which is the breadth, by 27 inches, which is the length, and it produceth 378. Having so done, divide this 378 by 144 and it will give the quotient $2\frac{90}{144}$, which is 2 feet and 90 square inches, or 90 parts of 144, which is very near two thirds of a foot, so that there is 2 feet and near two thirds in this flat board, whose breadth is 14 inches and length 27 inches. Now the reason of your multiplying the breadth by the length is to produce the quantity of inches, which is 378. If you strike 14 lines in the breadth and 27 lines in the length will produce 378 single parts, and why you divide it by 144 is that 12 lines being struck each way will be 144, and 12 inches being a foot and divided 12 times will produce the same for a divisor. So that, these reasons being observed, you are master of all flat measure, the multiplication, bringing all into inches, the division into feet square, which was required to be known. In like manner is all flat measure wrought, whatsoever.

To measure any square timber by arithmetic, as suppose a piece of timber which is 15 inches broad and 18 inches thick, and 5 feet long. I desire to know the square feet of timber contained in this piece. Work thus. Multiply the side 15 by the thickness 18, and that produceth 270. Having so done, turn the 5 feet which is the length of the piece into inches, by multiplying it by 12. It will give 60 inches long. Now have you the length and breadth in inches. Then multiply 270, your square of the piece, by the length 60, and it yields 16200. Having so done, divide this 16200 by 1728 and the quotient will be $9\frac{648}{1728}$ feet, the true content of the piece above mentioned. The reason why it is thus multiplied is to bring the whole piece into square inches, and why the divisor is 1728 is by reason so many square inches is in a foot square, as in the last chapter appears. For 12 inches flat has 144 upon one side, therefore it must be 12 times 144, if it be a foot thick as well as breadth, which was required to be made manifest. In this manner is all timber measured which hath breadth and thickness, but you will find some pieces of many sides equal or unequal, and round, all of which you shall find in their due places as we proceed. And by the way, remember that every 12 inch square of board hath 144 inches, but every foot square of timber hath 1728 inches, being 12 times so much by reason of the thickness.

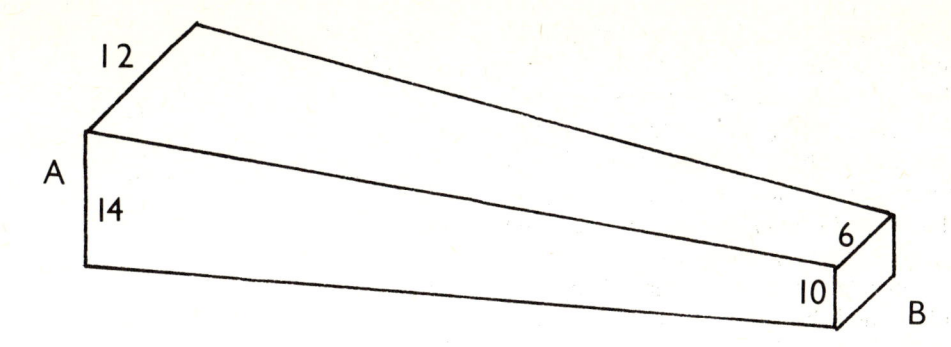

p32 To measure a piece whose sides and ends are unequal, as suppose a piece whose end A is 14 inches broad and 12 inches thick, and his end B 10 inches broad and 6 inches thick, being 24 inches long. I desire to know the content in square feet. This piece being so irregular, both sides and ends, you must find a mean diameter[1] for its square, which do by taking the half of 10 and 14, which is 12, and the half of 12 and 6 which is 9, then multiply 12 by 9 and it produceth 108. Having so done, take the difference which is between 14 and 10, which is 4, and the half of the difference of the other side, which is 12, and 6, whose half is 3, and multiply them one into the other, saying 3 times 4 makes 12, then divide 12 by 6 and it produceth 2 in the quotient, and add this 2 to the 108. It makes 110. Out of this 110 extract the square root, and the quotient will be $10\frac{1}{2}$ inches, very near, for your mean diameter of this piece. Now multiply $10\frac{1}{2}$ by $10\frac{1}{2}$. The product will be 110. This 110 multiply by 24 inches, being the length of the piece, and it produceth 2640. Then divide this 2640 by 1728 and the quotient will be $1\frac{912}{1728}$ feet. Why I take these means, observe that when you had taken the difference between the ends there remains a pyramid on the side, whose content was 2 inches, which I added to the former, and why divided by 6, is by reason every pyramid is the 6th part of a cube whose bases are equal, as you will find in its due place. And therefore I shall say no more, having showed you the hardest way of timber measure, which being observed you may resolve any question whatsoever.

To find how many inches in length will make a foot of square timber. By this method is graduated the carpenter's ruler, which discovers how many inches in length makes a foot of any square measure, as for example I would know how many inches in length will make a foot of a piece being 18 inches square. You must multiply 18 by 18 and it produceth 324. This 324 must be the divisor, and see how many times it can be had in 1728, which are the inches contained in a cube foot of timber, and your quotient will prove $5\frac{108}{324}$, which is 5 inches and a third of an inch in length, which makes a foot of square timber, and thus is the whole rule made. Take another example. I would know how many inches in length would make a foot at 12 inches square. You must work as above said, multiplying 12 by 12, makes 144 for your divisor. Now see how many times 144 can be had in 1728, and your quotient will be 12, which is the true length of a foot of timber whose sides are 12 inches each way. By these I suppose you well understand the making or graduating the common measure or carpenter's rule, both for board and timber, always remembering that the square inches of every foot of board, being 12 times divided and is 144, and then by consequence every piece of timber which is 12 inches thick must be 12 times 144, which is 1728, as is demonstrated in the first chapter in this folio. Now having showed you, this method, and which well observed needs no more examples.

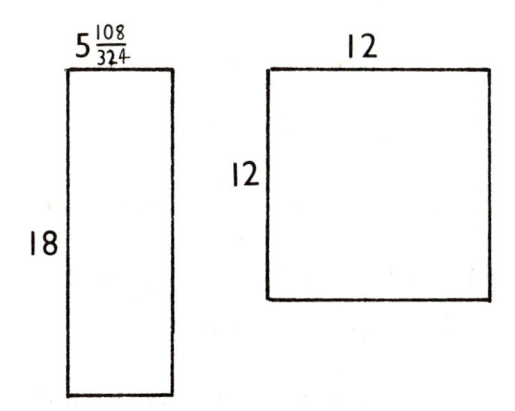

To measure girt timber, globes, and weight of shot by arithmetic.

p33 To measure round timber, called girt measure, as suppose a piece of round timber whose diameter is 15 inches from A to B, and 16 feet long from C to D. I desire to know the square feet contained in such a piece. Work thus. Multiply your diameter 15 by 15, and it produceth 225. Having so done, multiply 225 by 11. It yields 2475, and then divide this 2475 by 14 and the quotient is $176\frac{11}{14}$, and now have you the square inches of the diameter 15. Having so done, turn your length which is 16 feet into inches by multiplying it by 12 inches, and it produceth 192. Having so done, multiply your breadth of inches,

---

1. *Diameter.* Used in its older sense of 'The transverse measurement of any geometrical figure or body; width, thickness' (*Oxford English Dictionary*). This passage is also somewhat obscure. It is not clear whether the ends of the figure are parallel or not, and it seems pointless to find the square root of 110 only to square it immediately afterwards.

which was $176\frac{11}{14}$, by your length 192 inches, and the product will be 33792, which if you divide by 1728 the quotient will prove $19\frac{960}{1728}$ feet, which is something better than 19 feet and a half, and thus is your piece measured, and all other round timber by this method. The reason why you multiplied your diameter, after it was squared, by 11, and divide it by 14, is that every circle is eleven fourteenths of a square[1], and so is in proportion to it as 11 is to 14, and why your divisor is 1728, by reason that number of inches is in a foot cubic, so that the second multiplication brought your circle into a square equal to that circle, and your third multiplication brought both length and breadth into square inches, and lastly your division brought out the square feet, $19\frac{960}{1728}$, which was required to be done.

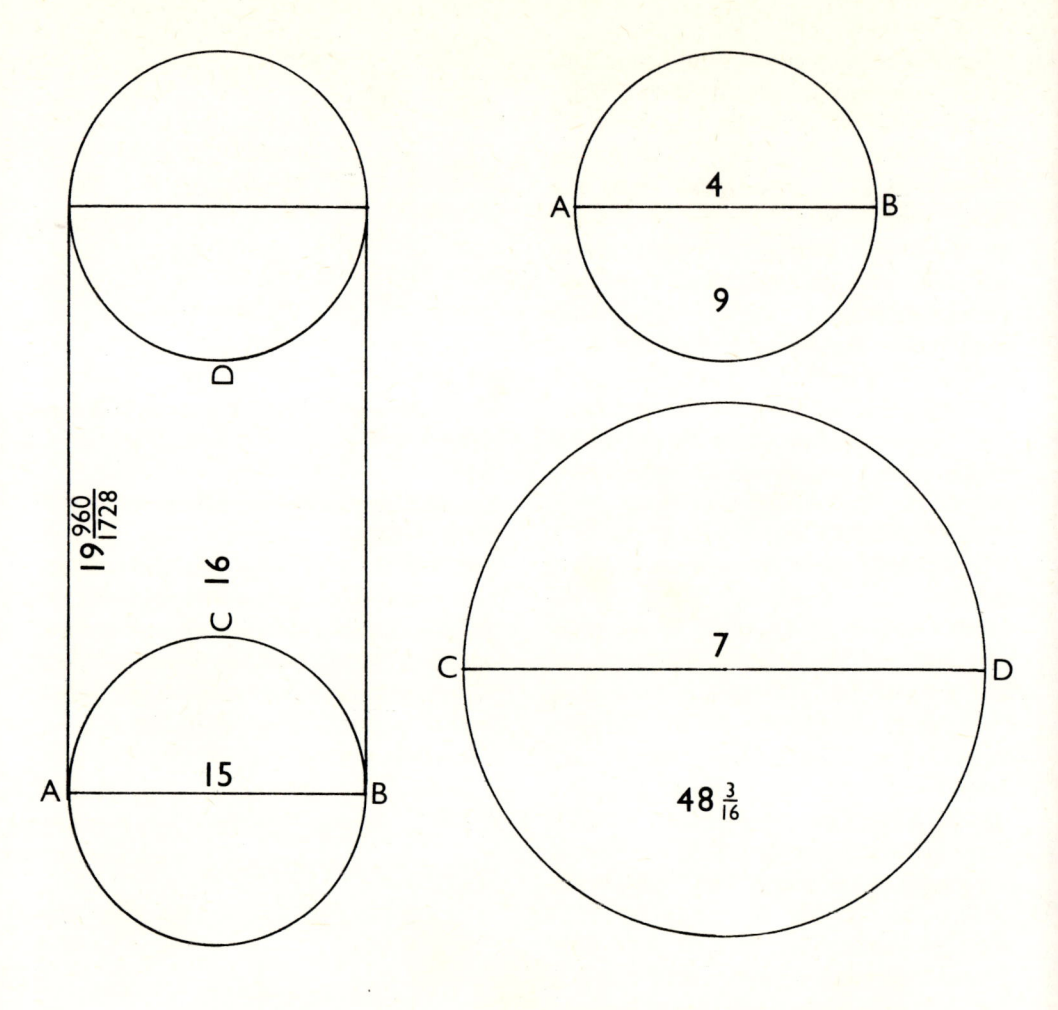

To measure any globe. I suppose a globe whose diameter from A to B is 45 inches. I would know how many square inches is contained in such a figure. Work thus. Multiply your diameter 45 by 45, and it produceth 2025, then multiply this 2025 by 45, and it produceth 91125. Now multiply this by 11. It yields 1002375, which divide by 21. The quotient will be $47732\frac{3}{21}$ square inches, in this globe whose diameter was 45 inches. And if you would know the square feet, divide this 47732 by 1728, and the quotient will be $27\frac{1366}{1728}$ feet. And thus are all globical bodies measured. The reason of your multiplying 45 the diameter by 45 twice is to cube it, and why you multiply that product by 11, and divide that by 21, is because every globe bears proportion to a cube as 11 is to 21[2], for if your globe figure were square and of the same diameter it would waste from 21 to 11 by cutting off the corners to bring it directly round every way. There is another way to work, by multiplying the diameter of the circumference cubically, and divide the product by 6, and the quotient will give the same number directly as the former. And thus I shall end this work, hoping that by these two ways you do well understand the measuring of any globical figure, always remembering that they do bear in proportion unto their squares as 11 is to 21, which is the reason of your work.

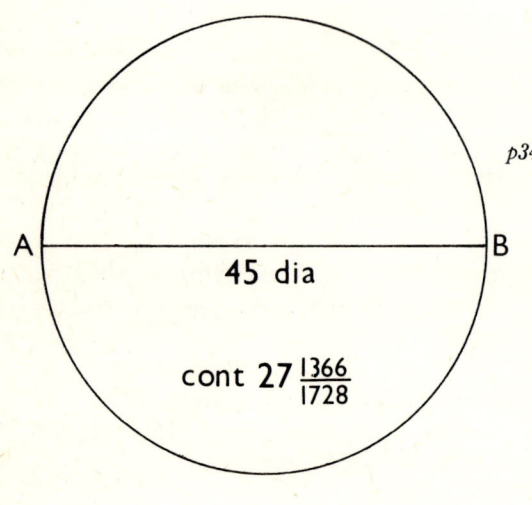

p34 To know the weight of any shot by proportion. Suppose a shot of 4 inches diameter from A to B. I would know how many pounds such a shot weigheth. You must work thus. Multiply your diameter 4 by 4, which is 16, and that product 16 by 4 and it will be 64. Having cubed your diameter by twice multiplying it, you must add the same number of your cube, and one quarter more. As 64 was your cubed product, add 64 to 64 and it makes 128, and then and $\frac{1}{4}$ of 64 which is 16 unto 128 and it makes 144. This 144 is the true content of ounces in such a shot, which divide by 16 and it will produce in the quotient 9 pounds, and thus have you your content, in pounds, of a shot whose diameter is 4 inches. Again, suppose another shot of 7 inches diameter from C to D. I work as above, saying 7 times 7 is 49, and then multiply 49 by 7 again and it produceth 343, which is the cube of 7. Having so done, I double the number 343, which is 686. Unto this 686 I add $\frac{1}{4}$ part of the cubed number of the diameter, which is 85, under 686, and it produceth 771, the content of ounces of a shot of 7 inches diameter, which, divided by 16, gives the quotient 48 pounds

1. *Every circle is eleven fourteenths of a square.* That is, area of a circle = $\pi r^2 = \frac{22}{7} \times (\frac{1}{2}d)^2$
$= \frac{22}{7} \times \frac{d^2}{4} = \frac{11}{14} \times d^2$

2. *As 11 is to 21.* This would appear to be a mistake, since it should be 11 to 28, but the calculation is continued using the figure of 21.

and $\frac{3}{16}$ of a pound, which is three ounces. And now having showed you two demonstrations, I shall not say any more as to this work, it being all performed by this method, in shot of any sizes whatsoever. The reason of the work is plain and needs no more interpretation and so pass unto my other work, hinting only upon every [blank] we may not lose too much time.

To know the content of shot, one from another. In the last you were instructed how to find the weight of any shot by its diameter. In this you shall find them from one another, as suppose a shot of 4 inches diameter weighs 9 pounds. I demand how much one of 7 inches diameter shall weigh. You must work thus. Cube your 4 inches diameter, by saying 4 times 4 is 16, and 4 times 16 is 64. Having done with that, cube your diameter of the shot required, which is 7 inches, by saying 7 times 7 is 49, and 7 times 49 is 343. Having cubed both the shot's diameters, work thus by the rule of three, saying if 64 gives 9 pounds, what shall 343 give? So that if you multiply 343 by 9, and its product is 3087, which divided by 64. The quotient is $48\frac{15}{64}$ pounds, the weight of the shot whose diameter is 7 inches. In observing this work, you may resolve all questions both of shot and powder it being all wrought by this proportion. And the allowance — powder, is common to allow $3\frac{1}{2}$ ounces to every hundredweight the gun weighs which shoots such shot, which, if you well observe, all art of gunnery is brought to pass by this method of proportions. And therefore I leave you to your own practice in whatsoever you please to make trial of.

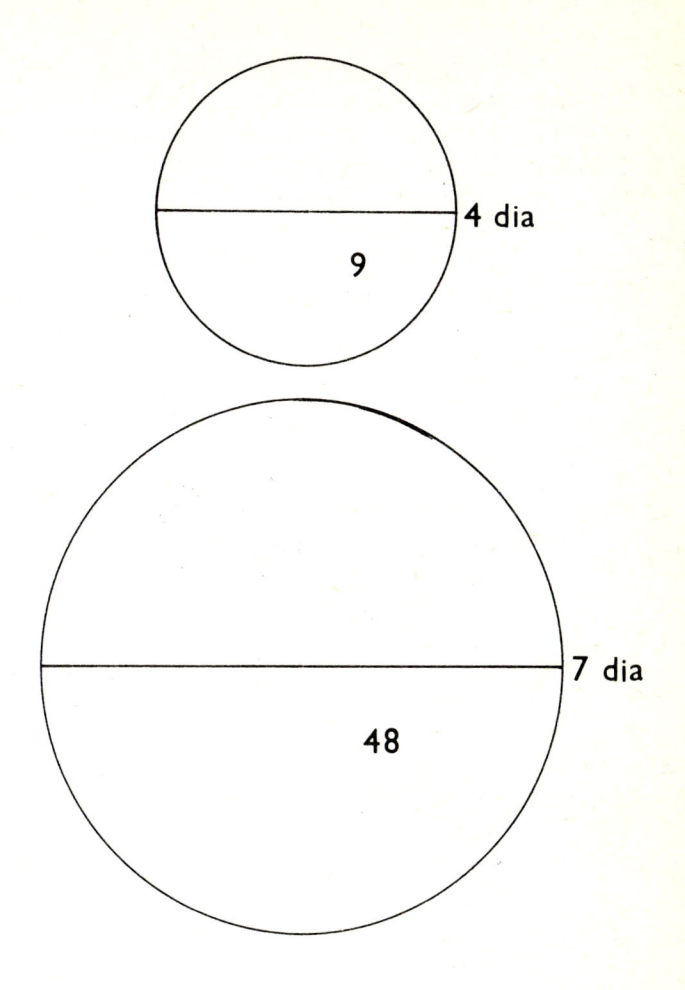

To measure by proportion the length or scantling of any ship's keel, or timbers one from another, to find the tonnage of any ship, and to gauge any round or square vessel.

p35 To measure the scantlings of ship's timbers by proportion, as suppose a ship of 400 tons, have a floor timber of 12 inches[3]. I desire to know the bigness of a ship's floor timber of 800 tons. You must work thus. Cube the square of your floor timber, which is 12 inches, by saying 12 times 12 is 144, and 12 times 144 is 1728. Having cubed your floor timber by your twice multiplying it, and that you would have the ship as big again, you multiply your 1728 by 2 and its product is 3456, which extract by the cube root, and your quotient will prove $15\frac{81}{401}$ inches, for the scantlings of a floor timber for a ship double to the former. But perhaps sometimes your proportion will not fall out double or treble so big as the former. If not, work thus, as suppose

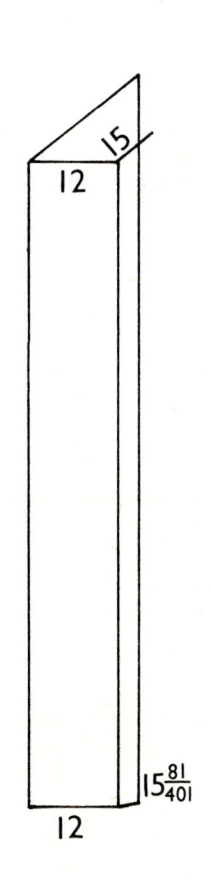

3. It is not clear how the diagram is intended to illustrate the text.

a ship's beam which is 12 inches square, whose burden is 500 tons. I would know the bigness of a ship's beam whose of 1050 tons. Work thus. Cube your 12 inches as afore taught, saying 12 times 12 is 144, and 12 times 144 is 1728. Having so done, say if 500 gives a cube of 1728, what shall 1050 give? Which by multiplying the second and third number together and it produceth 1814400, and divide by the first number 500 and your quotient will prove thus — $3628\frac{400}{525}$. Out of this number 3628 extract the cube root, and your quotient will prove $15\frac{253}{261}$ inches. Having now showed you two ways of measuring the scantlings by proportion, I shall need say no more. You well understanding, this method will answer whatever you desire, nor can any proportion hold except it is wrought by this manner of labour.

To make ships of like shapes and dimensions in every part, as suppose a ship's keel is 100 feet long whose burden is 544 tons. I desire to know the length of a ship's keel which is 978 tons. Work thus. Cube your 100 by saying 100 times 100 produceth 10000, and 100 times 10000 is 1000000. Having so done, work by the rule of three, saying if 544 gives 1000000, what shall 978 give? Which, by multiplying the second number by the third, and divide by the first, your quotient will prove thus 1797794, out of which extract the cube root, and the quotient will prove 121 feet and something better, which is the true length for a keel of a ship which is 978 tons. Again, suppose the same ship's breadth, of 544 tons, be 31 feet. I would know the breadth of a ship of 978 tons. Work as before, cubing 31 by saying 31 times 31 is 961, and 31 times 961 is 29791. Having so done, work by the rule of three, saying if 544 gives 29791, what shall 978 give? It will produce 53558, out of which extract the cube root, and your quotient will prove 37 feet and something better. So that by these two examples you may perceive how to proportion every part about the whole ship, making the great like unto the smaller, or the smaller as the great. But perhaps some will say this will produce shapes alike[1], but where are such shapes as are or may be called fit to be wrought, but my answer will be what my judgement leads me to believe, in another place, where the discourse is more suitable upon another subject of the like nature, and so pass to the next proposition.

p36 To measure the tonnage of any ship whatsoever[2]. In this matter I shall choose custom rather than truth, for ship's very full and broad will carry more than sharp or mean of like length, yet by reason sharp ships take up as much materials and labour, it is fit the builder should be paid the same rate for his works of equal length and breadth at the beam, although the owner of the ship desire her meaner, for some quality which best suits their occasion, or else I would show better, and more truer ways for measuring any burdens, as the ship would just carry by the true content of her body and her own weight. But custom taking place, here shall be set down the usual rule, as suppose a ship 100 feet long by the keel, meaning from the back of the stern post to the touch of the stem, and 31 broad in the greatest breadth. The half breadth must stand for her depth to measure, by which is $15\frac{1}{2}$. Work thus. Multiply your 100 feet length by 31 your breadth, and it produceth 3100. This 3100 multiply by $15\frac{1}{2}$, your depth, and it produceth 48050, which divide by 94. Your quotient will prove $511\frac{12}{94}$ tons. And so much would the shipwright be paid for a ship of these dimensions. It's needless to set down any more examples. This work observed is the way of measuring any tonnage according to custom, as you have been told, not by any truth, which you shall discover when you come unto your other work,[3] where we measure the whole ship's body and show the reason of her swimming or sinking by her own weight, and also her lading.

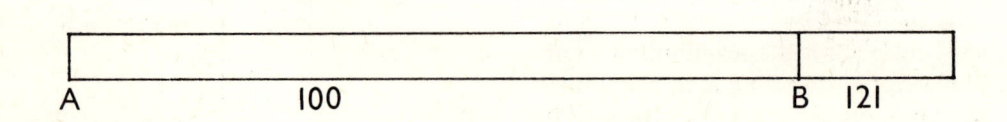

---

1. *Shapes Alike.* That is, what is the ideal shape for a ship, which can be scaled up or down according to need, a question which Deane intends to answer later. An example of how the same forms were used for ships of different sizes and rôles.
2. The system of tonnage measurement outlined here was standard until 1677, when the actual measurement of the keel was replaced by a theoretical figure produced by subtracting three-fifths of the beam from the gun deck length. Neither calculation was of any value in measuring the actual displacement, as Deane is aware.
3. The *other work* mentioned is on pages 69 and 70 of the original.

To know the content of gallons in any vessel round or square. Suppose a square vessel from A to B 20 inches, and from A to E 25 inches deep. I desire to know the content in wine gallons. Work thus. Multiply your breadth 20 inches, and it produceth 400. This 400 multiply by your depth 25 inches, and your product is 10000, which divide by 231 and your quotient will prove $45\frac{105}{231}$ gallons. But if you would have it ale gallons let your divisor be 282. Again, suppose a butt whose diameter at the bung is 32 inches from A to B, and diameter at the head 30 inches from D to E, and length from I to K 50 inches. I desire to know the content in wine gallons. Work thus. To find the mean diameter take two thirds of the bung which is 32, and it will be $21\frac{3}{4}$, and one third of the diameter at the head, which is 30 inches, and it will be 10 inches. Having so done, add your $21\frac{3}{4}$ to 10 and it makes $31\frac{3}{4}$ inches, shall be the mean diameter between the bung and the head. Having found your diameter, square it, saying $31\frac{3}{4}$ by $31\frac{3}{4}$, and it yields 982, which multiply by 11 and it produceth 10802, and this 10802 divided by 14 your quotient will be 771, for your square diameter, which multiply by the length which was 50 inches and the product will be 38550, and divided by 231, the quotient is 166 gallons, the content of such a cask in wine measure. By these two vessels, square and round, you may with your own practice measure all others at your pleasure.

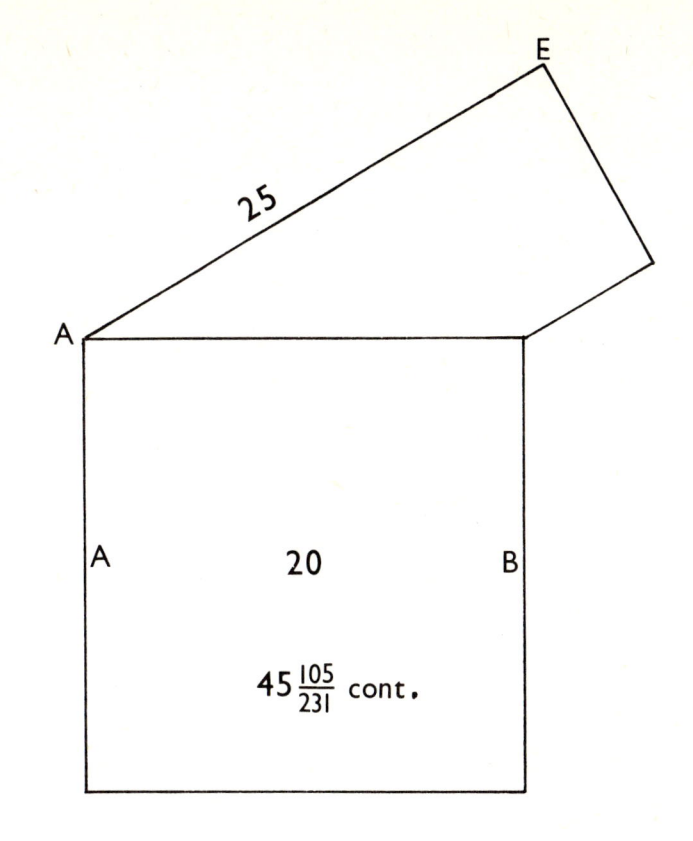

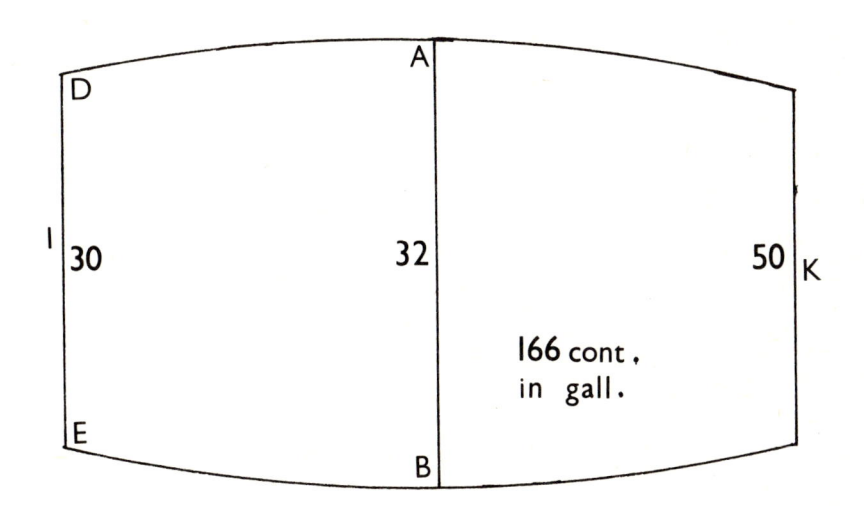

Part 2 Hull Design

The base or bottom of a ship's keel demonstrated, being the first work in drawing a ship's draught, and the only line whereon all the rest depend.

p37 The first stroke that is struck in this work is the line of the keel, and is begun on the plate with the striking of a straight line, at such a distance on your plate, board, or paper as to have room on the board for your half breadth of the ship, and to have room for head and stern as to the length of each part, and therefore it is to be considered how big your scale is, and whether your board or paper be big enough to hold all the matter you intend for your design, and if your scale is too big, then you must either make a smaller, or enlarge your plate, board or paper on which you intend to draw. And having thus fixed your scale, proceed to draw a line in black lead from one end to the other of the ship. And by the way I must tell you, for your more curious drawing without blemish, draw all your whole draught in every line with black lead, and then when you have done, you may ink it altogether, and then rub over all the draught with the crumb of white bread and it will take out all the black lead, and by this means the draught will be very clean, and be complete as you desire. Which having advised you, I shall proceed to the work, drawing the straight line long enough for the ship's keel, and also the rake afore and abaft, by reason it is to serve for the bottom of the keel and for a middle line for half the breadth of the ship, which you will find in its due order. For I do intend to begin with the meanest thing and so leave nothing unfolded which may advance anything to the meanest capacity, until the whole work be ended, and therefore desire the suspension of anyone's judgement until they have viewed the whole work, and where they declare the defect of what they may perhaps find, then to take the pains to mend it, by which means, when I may give it the view of what they have found out, I may be convinced, and so agree, that, as it ever has been, by falsehood truth appears. And having thus done, do proceed to draw the line for the bottom of the ship's keel in black lead, from one end of the ship to the other, and also for the rakes both of the stem and stern post, which I call the bottom of the keel, and have marked it with A and B as is shown on the other side. Which line must consist in length 152 feet, that is to say 120 feet for the keel and 26 feet and a half for the rake of the stem, and 5 feet and a half for the rake of the stern post. And thus have I done with the first line, namely the bottom line of the keel, and shall refer you more clearly to the other part, as on the other side is more plainly shown. Having done with the bottom line of the keel, and concluding the ship to be 120 feet in length, I have thought it convenient to inform you of the breadth of the body, she receiving many of the scantlings from that part, and do for breadth direct 36 feet from outside to outside of the timber in the midships. To let you know from whence I derive my breadth, I take $\frac{3}{10}$ of the length of the keel, being 120 feet, and find it to be as appears, 36 feet for breadth, which will be breadth enough for any man-of-war, and to make a stiff ship, but for a merchantman one third of the length of the keel is better for the burthen, for the less water the body opens it is the better for the motion, if she be capable of bearing a stiff sail. Which breadth now being concluded, I proceed to the former work again, which is as followeth.

A ———————————————————— B

The length of the keel demonstrated, and bound by two perpendiculars.

p39 This other problem[1] shows the bounding of the line for the true length of the bottom of the ship's keel, at which is raised two perpendiculars, the one for the stern post and the other at the stem, by which means the length of the ship is now discovered to be 120 feet as intended. And lest any that view this should not readily know what is meant by the perpendiculars, I here make it clear to their advantage. By perpendiculars is meant those two pricked lines which are upright from the line of the keel, and are the bounds of the true length of the keel, and are for direction for the rake afore and abaft, and are geometrically raised on the other side. As suppose the keel line A and B, on which you would raise those upright lines, then set one leg of your compasses at the place you design it to stand. Having thus done, open your compasses to any distance, imagining it to E and D. Having thus set off from

1. *Problem.* Used in the sense of 'A proposition in which something is required to be done' (*OED*).

HULL DESIGN

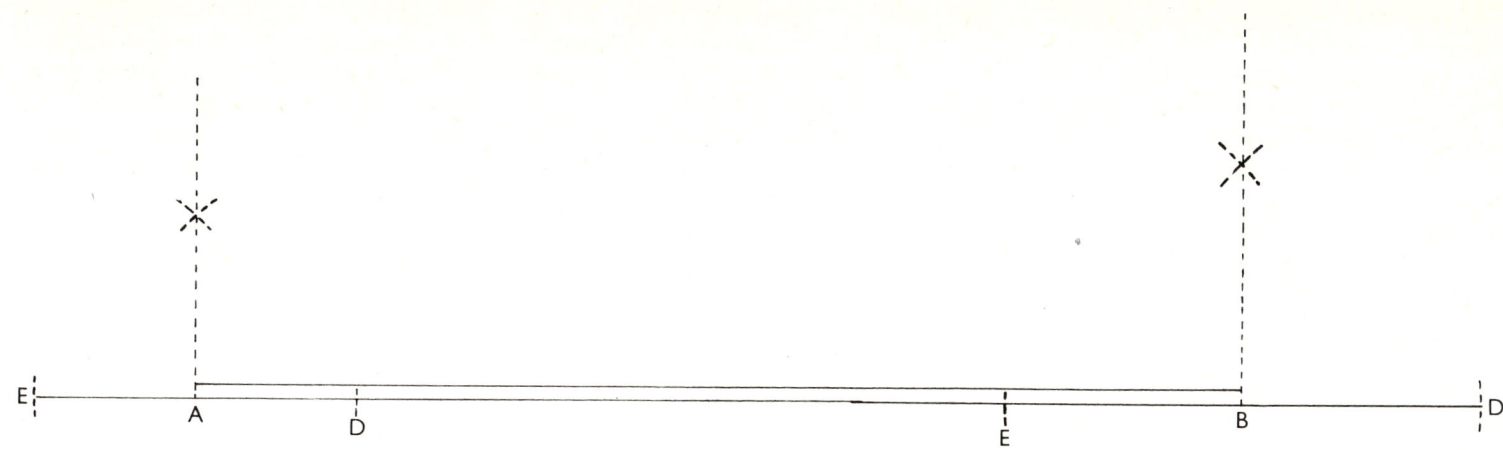

each side of your line A and B and E D, then open your compasses yet wider and strike an arc by setting one leg of your compass in D, and, having done, set the other in E and strike the like arc. With which when you have done, lay a ruler on the middle prick between E and D at A, and where the arc cuts, and strike a white line with the leg of your compasses. Then have you done with a perpendicular, which I think most know that will delight in this work. However it is not improper to effect my purpose, although few that draw which will not make a square save this points though a mechanic way,[2] but it most fit for that, which is all I shall hint as a perpendicular, which is often in use before we end our works. And therefore do not question the industry of him that differs to be a proficient in this.[3] Affairs will carry this word and the natural perpendicular in his memory, as he that learns to read must remember his letters. And therefore shall get to my other matter, having done with the bottom line of the keel, and bounded it with two perpendiculars, so do proceed to the next plate in due order, the just length of the keel being found to be 120 feet.

The length of the keel, rake of the stem and stern post demonstrated.

41 To this work on the other side, which is the bottom of the keel line, and also shown how to raise the perpendiculars, the one from the end of the keel whence the stem is to be joined, the other for the stern post to be joined, also is added, namely, the distance from B to C, 27 feet and a half for the rake of the stem, and is set off parallel to the other perpendicular B, by which means the extent of the rake of the stem is known, and so bounded with those two pricked lines. It remains now to tell you from whence that rake of 27 feet is derived, and it will be found to be $\frac{3}{4}$ of the main breadth, 36 feet, which will be very sufficient for a ship of war of any rank whatsoever. So have we done with the perpendicular of the stem, and the true distance from the perpendicular of the length of the keel, which are as above mentioned. From C to B is 27 feet or very near, is a good proportion, and so shall I say no more as to that. It remains, now you have the rake of the stem and how it is derived, that the stern post be also shown, which is as followeth. Take from the rake of the stern post near $\frac{11}{12}$ parts of the $\frac{1}{6}$ of the breadth, which is 36 feet, and you will find it to be five feet and a half, for the $\frac{1}{6}$ of 36 feet is 6 feet, and $\frac{11}{12}$ of 6 feet is 5

2. *Few will draw ..... mechanic way.* Deane appears to be referring to the drawing of a perpendicular by the use of a set square or similar instrument, in contrast to the compasses which he uses.

3. *Do not question ..... proficient in this.* Perhaps meaning that the other method, by the use of the set square, is also acceptable.

53

feet and a half, which is a good rake for the stern post, which is from A to D 5 feet and a half. Now have you the length of the keel, and the rake of the stem and stern post, which are both set off parallel to the other perpendiculars of the length of the keel, in their due order. And having done with the length of the keel, the best depth must be sought for, which having well considered, I take from the scale 18 inches for its depth, and having thus done, I set it with my compasses from the bottom line of the keel, A and B, and then strike another line parallel to the former, which is EF, and now have you the depth of the keel, which is one foot and a half, I mean with the false keel so well as the main keel. Which false keel I order to be 3 inches, so that the main keel need be but 15 inches deep, but before I pass by this I shall give some hint for the depth of the keel, and a proportion for it. First, for its whole depth I allow $\frac{1}{2}$ an inch in depth to every foot the ship is broad, and then 36 feet gives 18 inches, though I cannot but confess more keel would do no harm, but much good, in sailing to windward, for nothing has less motion to the leeward than a flat thing and deep against the water, by which it would cling a ship well to windward, but here is the danger, that if it should hap that the ship should run aground by any accident, which all ships in some measures are liable to, then by extreme depth of the keel the ship would be so over-heel her bilge that with her own weight, and weight of her guns and the like, that she were utterly ruined, so that our aim is to reconcile all inconveniences to the best advantage as well aground as afloat. More reasons may be laid down, but this one is sufficient to any who knows a ship. Now for the false keel. The main end[1] is to lay it with tar and hair, and spike it to the main keel's bottom, for the preservation of the main keel if the ship should proceed to the South Seas, where the worms are so prevalent in eating of all ships underwater. Which is instead of sheathing[2], for the sides of the keel can be sheathed in a dock, but the bottom of it cannot be come at, without great charge of getting away the blocks she lies on and the like. There be other reasons for it, but this one is enough to acquaint the use of them, which false keel and main keel being 18 inches as is before shown, a good proportion for a ship of these dimensions, and will lie aground without danger. There is two other ways[3] for the rake of the stern post. Take 3 inches to one foot from the bottom of the keel to the main transom, or one quarter part of its length, and it will be a good rake also, so well as the former, and will prove near five feet and a half. Now having three ways of doing it, I leave you to your choice, and so proceed to the next plate, showing the sweeping out the stem and also the length of both the stern post and the stem, as on the other side is shown.

The sweep of the stem and stern post demonstrated, with the bigness of the same.

p43 Having shown on the foregoing plates the length and depth of the keel, with the rake of the stem and stern post set out in its due place, I come now to the sweeping out the stem, which is done thus. Take off from your scale 27 feet, which is the true rake of the stem, and set one leg of your compasses in the perpendicular of the stem E, and let it touch at the bottom line of the keel, just at the perpendicular which bounds the length of the keel, and then sweep it from A to B, and, when you have so done, that line is the outside of the stem. And then, when you have done that, sweep another of the same form[4] for the inside of the stem, and then have you the length of the stem. And for its thickness and depth[5], let it be $\frac{8}{9}$ of the breadth of the keel, and then it will appear to be 16 inches, which is enough, only at the head keep it the whole depth of your keel, which is two inches more, and thus have you the true dimensions of the stem and its derivation. It remains now to show the length and bigness of the stern post, which will be found thus. Take from your scale $\frac{2}{3}$ of the breadth, 36 feet, and abate half a foot in that number, and you will find it to be 23 feet and a half, which shall be the true length of your stern post from the bottom of the keel to the main transom or height of breadth, which place from C to D and strike it diagonal to the two perpendiculars, which line makes the outside of the post, on which the rudder is fayed[6] against. And now have you the length of the stern post to the transom. It remains now to show the bigness fore and aft and thwartships, which at the head ought to be $\frac{2}{3}$ more than the depth of the keel, which will then prove 30

1. *The main end is ......* . The usual reason given for the false keel was to protect the main keel against excessive damage if the ship ran aground.
2. *Sheathing.* At this period it usually meant covering the underwater body with tar and hair, and covering that with a thin layer of planks, to prevent shipworm. Around this time lead sheathing was being tried experimentally, and unsuccessfully.
3. *Two other ways.* But only one is mentioned.
4. *Sweep another of the same form.* This suggests that the inside of the stem post should be of the same radius as the outside, but it is more likely that it would have the same centre but a smaller radius.
5. *Thickness and depth.* It is not clear what is meant here by depth, since it has already been drawn on the plate. Thickness refers to the width of the stem post. The keel normally narrowed towards the ends, and presumably it was 16 inches wide at the forward end where it joined the stem post.
6. *Fayed.* To join two pieces of timber together.

inches, which shall be its thickness fore and aft, and also thwartships. For the lower end, it is the just bigness of the keel athwart, and fore and aft so big as your piece can afford, by reason it will make a good hold for the hooding ends[7] to fasten to. And for its shape, line it by two straight lines, on the back post on $\frac{2}{5}$ of its breadth at the head, which will be 12 inches, and then if occasion serve a false post may be brought to it. And for the shape in the inside line, it as full as you can, that its branching may fill out as the ship's run grows thicker upwards, which experience in the work makes manifest. And thus have you both the length and bigness of the stem and stern post and their thicknesses and shapes, so shall we proceed to the other part of our work as on the other side.

Note: The shape of the sternpost as shown in the plate is not fully described in the text. Also shown on the plate is a double line at the top of the keel and the rear of the stem post. This probably represents the rabbet of the keel, a groove into which the edge of the planking was fitted. This double line is not repeated on all the following draughts, and it is not included on the final one. No dimensions are given for the height of the stem post, and this also varies from page to page. In general the drawings have been copied as in the originals, including the inconsistencies.

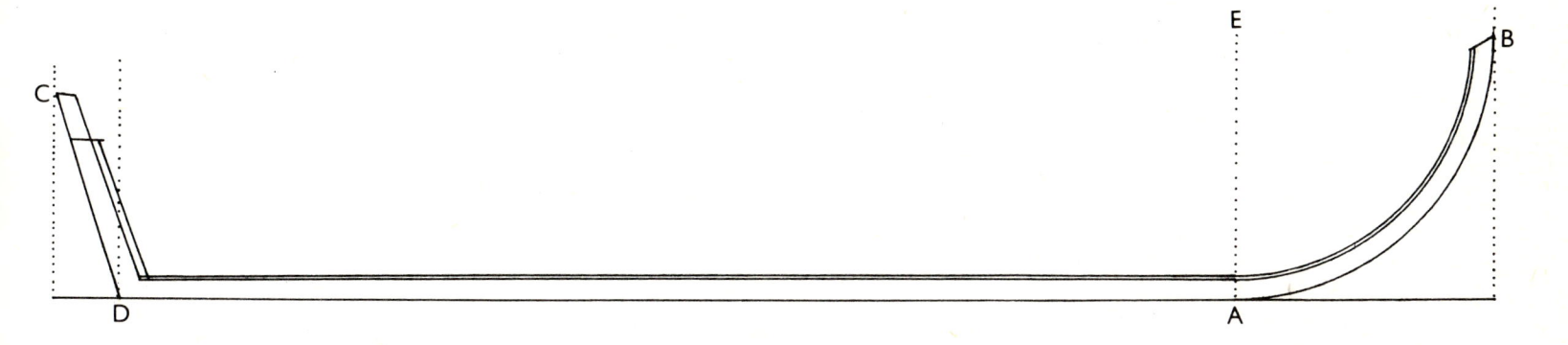

The water-line, or greatest depth of water the ship must draw, completely gunned, rigged, victualled, and stored.

p45 There is added to this problem on the other side only the water-line, which is of so great concernment that the whole good or bad quality of a ship depends on that design, and is, or ought to be, the principal line regarded in the whole art of a master shipwright. For to lay down this line as it ought to be is great worth and advantage on every man-of-war, which, if it had been thoroughly understood, we had never had such great mistakes in the navy as we have found, to the confounding of a vast treasure in the kingdom, by sometimes lengthening them, or by girdling[8] them of great thicknesses, at other times making them fuller or leaner bowed, and perchance furring out[9] or thickening the run[10] or the way, also aloft in the quarters, as the fault may be gathered from those who command them at sea. And perchance it falls out that the alteration is made by one who conceives not the tons, and then, instead of mending what was amiss, makes it rather worse in some other matter, as it is not very difficult to

7. *Hooding ends.* The stern post was thicker at its after end, so that the planks could make a smooth join with it.
8. *Girdling.* Adding strips of timber to the outside of the underwater body of a ship in order to make it float higher.
9. *Furring out.* Similar to girdling in purpose, but involving a more basic alteration. In this case the timbers were stripped from the underwater body and a piece of timber added to each frame, and the hull re-planked.
10. *Run.* According to Mainwaring, the run was the underwater part of the body as it narrowed towards the stern, while the way was the narrowing towards both ends.

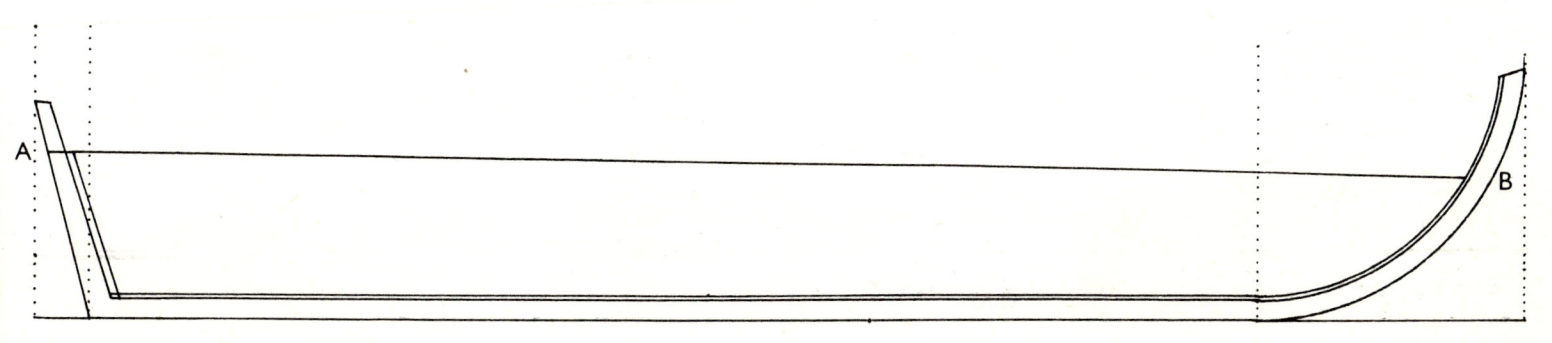

make a conceit[1] in that case. And in short all these mishaps afore mentioned and others, so well as the drawing 21 feet water when they intended but 18 feet, it flows only from the ignorance of not being assured, before they begin, what bigness of body, for ships of any dimensions, with such scantlings, will not fail of drawing such a quantity of water, light or when they be launched, and also so much and no more with so much provisions, guns, stores, weight of men, masts, rigging, and the like, being completed with everything in, fit for the sea, all of which is exactly to be known if the builder will take the pains, or his skill afford so much of that art. And now, having spoken of an assignment into the water and how much it concerns the builder, besides the good quality to know a just assignment, I shall speak something of its nature here, and at large in a proper place, and do assign, for a ship so large as this, but 17 feet water, and if it could be less it were better, to follow our enemies in shallow water, and it would be less danger amongst our own channels and sands. And why raising her so much is by reason of the height this ship's lowest guns ought to be — 4 feet 6 inches from the water at least, when all provisions is in, which causes her depth to be 15 feet 9 inches depth in hold, and 7 feet between decks, and 5 feet waist[2], with all convenient heights in the steerage[3], forecastle, coach[4], and the like, all which depth of body must be brought to windward in the sea[5], and therefore I conclude — 17 feet draught of water, a body well shaped and regular lines, will draw a ship of this bigness as well to windward as if it were more, and the power of the element of water will be sufficient to so much upper work as the ships I have built hath, and will manifest better than any form. But perhaps some would be ready to say it were good to make it out by demonstration what I have asserted, which I will at large, and only hint here. The reasonableness of this, being assured the ship shall draw so much water and no more, is the most likeliest way for well doing, for should you draw a draught so broad and full in every place that she would carry everything she ought at less than 17 feet water, as suppose 15 feet, then I say this ship's fullness would cause her to carry nigh 270 tons more to bring her to 17 feet, which, if it do, then I think it allowed for reason that if so much more weight of ballast be put in, or lading, or any weight else, that the ship opens more water by so much as she is bigger than her intended design, as supposing her body so big as to carry 270 tons more than what is needful, then I say she must make way through the just weight of water as the 270 tons amounts to, for the weight put into any ship is equal to the same weight of water the ship's body occupies, to an ounce or less. So that the case lies here in short, the master shipwright ought to design her body that with its own weight will carry 6 months provisions, which is sufficient for any man-of-war, which I suppose every seaman will allow to be enough, for many reasons which will be made plain in its due order. And also shall I lay you down both the scantlings for each piece of timber and plank, and the just weight of all things fit to put into her, and also to know, ere you begin, to assure yourself of the ship's burden shall be so great, and her own weight so much. And at last I conclude it by laying down this whole figure geometrically and by arithmetic, which is not fit here to trouble you with, until such time as you more understand the works. And therefore shall I conclude this line, laying it 17 feet abaft and at 15 feet 2 inches forward, which is 16 inches by the stern, for the ship's better steerage and quality, and have marked it with A and B, and call it the water-line, or line of assignment, when the ship hath all things in her ready for sailing, completed in every circumstance. Only, by the way, I had forgot to tell you this body must not be assigned less than to carry what is proposed, for if it do, the ship will sink deeper into the water, and then you are deceived with the laying of your guns, or perhaps the gun deck then under water, which I have seen several to the ruining of the ship. It is not in nature to make a deep ship, and not to put her down in the water to cause her to carry a stiff sail. And therefore I shall say no more in this place, only conclude with this, that a mean body, less than what is assigned, will go down deeper into the water than your intended purpose, with the weight which must be put into her, and then your guns lie too low.

And a fuller body will take more weight in by consequence, and will not move so fast with the same sail. So that we shall show you how to prevent both the extremes and make a mean, to answer everything you desire, and to be assured of your draught of water ere you begin your work, which done, you will build your ship to each point, and not have cause to repent of your labours, nor those who employ you. And so shall I cease with showing you the water-line at her greatest depth, which is marked with A and B.

---

1. *Conceit.* This word is obscure in the text, since it is near the end of the line, and part of it appears to have been missed out.
2. *Waist.* The uncovered space on the upper deck between the quarter deck and the forecastle.
3. *Steerage.* A cabin towards the after end of the upper deck, so called because it housed the whipstaff, which was used for steering.
4. *Coach.* A cabin built on the quarter deck, so called because it originally resembled a coach in shape.
5. *Must be brought to windward in the sea.* That is, the upper works must not be to high, or they will have an adverse effect on sailing qualities.

The shape and place of the lower wale demonstrated.

p47 And therefore we now proceed to the problem intended, namely the lower wale[6], on which depends the good or bad shape of any ship, for by this wale is all the others set, so that it is of absolute necessity this wale be well done. And therefore I shall declare you two ways of doing it, and prove the demonstration by arithmetic, by reason the great assurance of this proof lies in the artificial part, by which means you are assured ere your work proceed too far in error.

---

6. *Wale.* A piece of timber, thicker than the normal planking, which was intended to give extra strength to the hull.
7. *White straight line.* Possibly meaning a pencil line which is later erased from the completed draught.
8. *Screw up your bow.* The bow was shaped like an archer's bow, but with a peg which could be used to tighten the string, so as to set it to any curve. It could thus be used to construct a circle based on three known points.
9. *How to do it by arithmetic,* etc. This is based on Pythagoras' Theorem:

$$x^2 = (x-A)^2 + B^2$$
$$\therefore x = \frac{\frac{B^2}{A}+A}{2}$$

For the clear manifesting of this I take $\frac{3}{4}$ or something more from the height of the stern post, from the bottom of the keel to the A the upper edge of the transom, which $\frac{3}{4}$ I set from the bottom afore, and set it just perpendicular to the stem which touches at B. And having now assigned my height of the upper edge of my wale afore, so well as abaft, I strike the white straight line[7] from A to B, which line I find how much I am to hang my wale from a straight line in the midships from D to C, which I shall here direct two most excellent ways which will give you great content. First for my number of inches the wale shall hang, I consider how long the ship is from A to B, and find it to be 149 feet, or 49 yards. Then for so many yards the ship is long in that place I allow for every yard one inch and a quarter and a half quarter for its hanging. And then it will be 67 inches and $\frac{3}{8}$ of an inch, which I take off from my scale and set it downward from D to C. The other way is thus. I find my ship is 149. I multiply by four and divide that product by 9, and the quotient is the true hanging of the wale, so well as the former which is shown. For clear demonstration, suppose the 149 feet multiplied by 4. It yields 596, which 596 divide by 9, and the quotient is $66\frac{2}{9}$, which $66\frac{2}{9}$ inches is very near the 67 inches found out by the former way. And thus have you both ways to set off the hanging which is at

$499\frac{8}{12}$ FEET RADIUS OF THE LOWER WALE

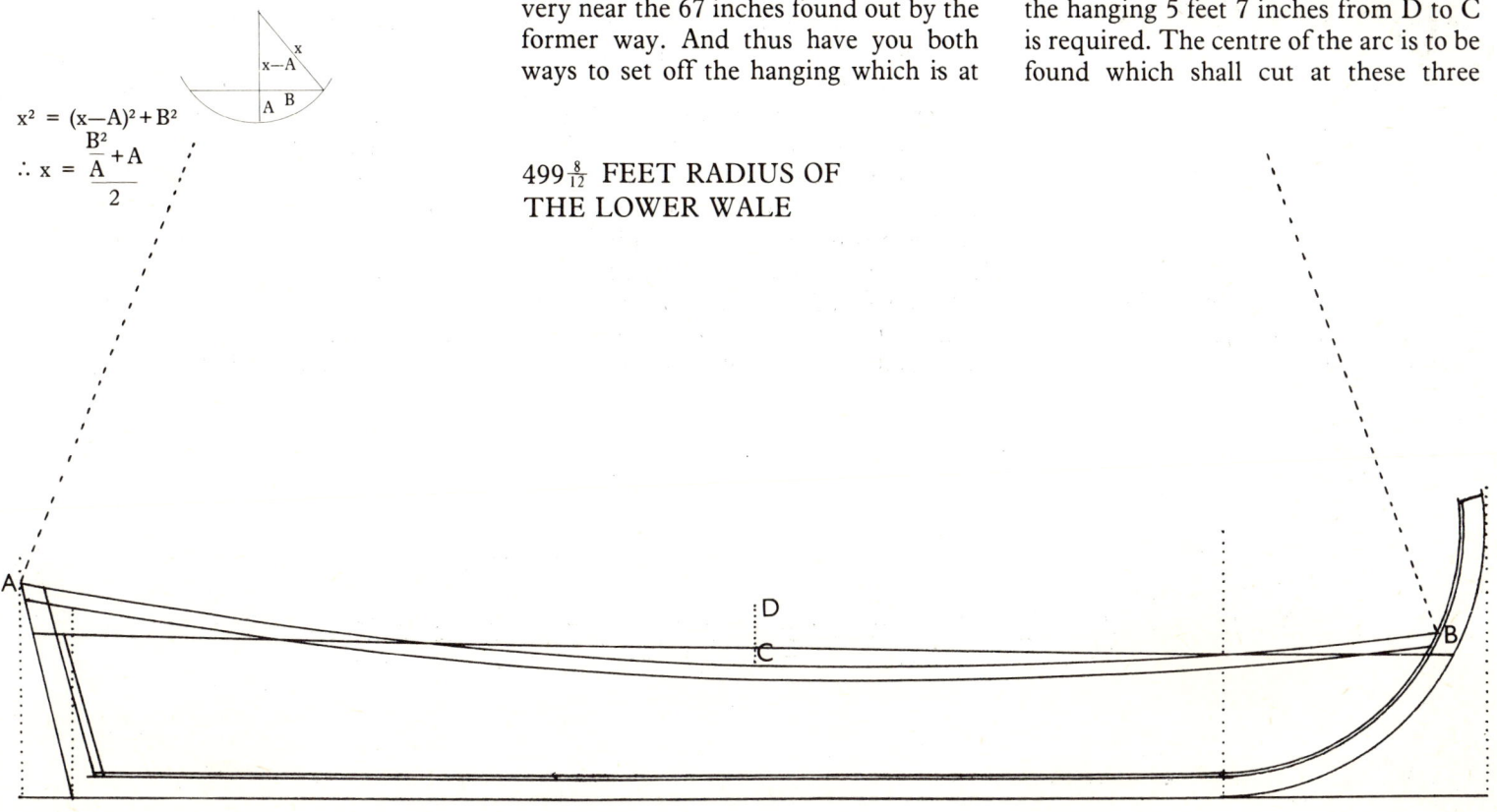

C. You must now to work the line mechanically. Screw up your bow[8] fitted for that purpose, which, when you have raised it to the point C, the line must be struck from ABC. And thus have you the upper edge of the lower wale. It remains now to derive the depth for that wale, which I do thus. Take $\frac{1}{2}$ an inch for each foot the ship is broad, which I take from my scale, 18 inches, the ship being 36 feet broad, and then with my compasses I set it downward and strike another line parallel to that of ABC, which line completes the lower wale, which line being well done you cannot err in your sheer or shape. But it so falls out that I must show you here how to do it by arithmetic[9], lest the young artist should hap of an ill bow, which may discourage his proceedings, and therefore we cannot well pass it by in this place, though indeed it was not in my thoughts to proceed in this till I had gone through the other, which I will not pass by for the clear understanding of the reader. And therefore I do proceed thus. You see the white line from A to B is a straight line and is in length 149 feet, and that the greatest hanging of the wale is in the middle from D to C. Now, forasmuch as the whole line AB is 149 feet, the half line at D is 74 feet and a half, which is just half of that angle ABC, and the hanging 5 feet 7 inches from D to C is required. The centre of the arc is to be found which shall cut at these three assigned places, and sweep the line out which is upper edge of the wale, which will prove thus. Multiply the half basis into inches, because the depending bears the same demonstration as the 74½ feet into 12 inches, it being one foot short, which being multiplied proves 894 inches, which is the length of 74 feet and ½, which you must square by multiplying it unto itself as thus — 799236. Now, as it is all brought into inches, divide it by the depending or depth of hanging, which is 67 inches, as thus — 11926. Unto this quotient must be added 67, and it will prove thus — 11993. Now have you the whole work in inches, which divide by 2 and the quotient will prove thus — 5996 inches, which number is the radius required, which I turn into feet by dividing it by 12, and then the quotient will be 499 feet and $\frac{8}{12}$, which number of 499 feet being taken off with your compasses from the scale, or if your compasses be not long enough, use a line or scale of 499 feet according to your scale, which if you sweep it out will cut the three assigned places and it will make that perfect line of ABC. And thus have you the whole work well observed, is to this art a great help, for the work above I shall make plainer, as thus. Had your basis been 74 feet, and your hanging 6 feet, which is like denomination, then it had been no more but square 74 by 74, divide it by 6, unto which product add 6, and the half of that would be the length of the radius required, but by reason one bears the name of inches, the other ought so also, for the truth of the work. And thus have I done with showing you how to make the lower wale as it ought to be. Which done, I proceed to my other matter in its due order.

The gun deck and upper deck demonstrated, together with the lower and upper counter.

1. *Same sweep or radius.* It is more likely that the radius should be reduced, and the same centre used.
2. *Knees.* Curved timbers linking the deck beams to the sides of the ship.

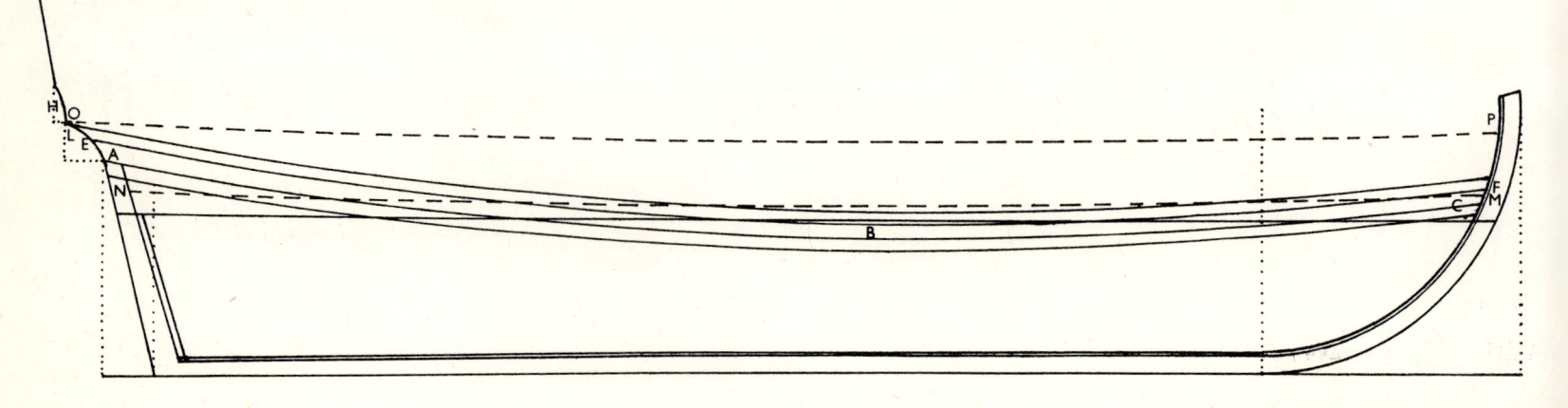

p49 This problem seems easy unto the former, for it hath only added to it one wale more, and the lower deck and upper deck lines, and the rake and height of the whole stern, as shall be shown in its due order. First, for the shape of the wale which is added it, is by the same sweep or radius[1] of the other on the other side, ABC, or by the mechanic way it is done by a bow whose lath is set by which you strike the same wale, only, as you strike the upper line first on the other wale, so now you shall strike the lowermost line, which is best for your direction. Now, to know what distance it shall be set from the upper edge of the lower wale, I conclude it thus. Look how many feet the ship is broad, and I take from it $\frac{2}{3}$ of so many inches, and it will prove 24 inches, which is $\frac{2}{3}$ of 36 feet, the breadth of the ship. Which 24 inches I take from my scale and set one leg of my compasses in the line ABC, and set that distance, 24 inches, in two places, the one afore and the other aft, and then strike it equal distance or parallel from ABC, and by the same radius or bow strike the line EF, the lower edge of the wale. And for the depth of the wale I take from the lower wale, which was 18 inches, $\frac{5}{6}$, and then shall I have 15 inches, which I take from my scale and set it equidistant as the former, and thus have you the line of the upper edge, and the wale completed. The reason for this distance of the wale are not absolute, but various, yet it is of most strength to the ship to lay them thus. They being so laid will be well fastened to the knees[2] in hold, being placed against them, and besides, the ports will cut but little of them abaft, and the lower edge will not come quite into the matter of the upper wale, and lastly, it cannot well lie in any place

58

where it will be so strong, and finely shaped, and is always practiced, though it is allowed some men will say they can place them higher, or lower, or make them more in number, which I grant may be, yet I advise the reader to keep in this way until he is sure of a better. Which is all as to the wales. Now we come to the lower- or gun deck line in its order. As to the gun deck line, I am ruled by my water-line, which I have taken such care to assign well, and, being assured of the ship's going no deeper into the water, I consider how high my guns ought to lie, and I find 4 feet 10 inches from the water a gallant height. Then do I set off in the midships from my water-line 18 inches, and afore and abaft 2 feet 10 inches, by which means my deck hangs 16 inches in the whole length. And having thus done, I set a bow into the three pricks which were set off from my water-line, and so strike the pricked line which is the gun deck, marked NM. But perhaps some will say, why may not this line be higher from the water-line, that so the guns may lie higher also, my answer is, the quality of a ship must be regarded so well as accommodation, and therefore if the deck were laid higher the ship would be crank[3], and besides, the ship would have more body of upper work to draw through the air, which nothing can do it, but by putting the ship deeper into the water. Which, if you should do, then were you to seek a new height for your guns, and uncertain of your whole work, and therefore conclude the deck well assigned to 17 feet water, unto which all things have a due correspondency. As to the other deck, it ought to be 7 feet from plank to plank between each line, and the same shape, which is the upper deck line and is marked OP. Now have you the decks. Place your ports ere you go any further, which make in width 2 feet, and in depth 2 feet 8 inches, and place the lower edge of them 2 feet four inches from the deck when it is planked, or 2 feet 8 inches in your draught, and let the number be 13[4], which is enough, and to keep the ship strong. Having thus done, your guns will lie as directed, which is a gallant height from the water. It remains now to show the rake of the counters, and their length, which I shall do thus. The lower counter marked L must rake so far aft as to hide the rudder head which comes just under it, and therefore I set it by a square 4 feet aft, and by a perpendicular 4 feet 3 inches. When I have done, sweep it by a 7 foot sweep, which I call the lower counter. And having done which I consider a small counter for carved work under the light of the great cabin windows, which I make one fourth of the lower counter, which will be $13\frac{1}{2}$ inches, the line H, and then set it perpendicular $\frac{3}{4}$ of the lower counter, and sweep it by the same sweep of the lower counter, which was 7 feet, and thus have you both the upper and lower counter assigned. Then for the upright of the stern, strike a straight line, and rake is something less than the stern post, by reason of your false post, which will cause the main one to rake less. All which observed will cause a very handsome stern, which in its due place will be showed, as well the breadth as the side which is only now shown, and so proceed to the other work in its due order.

The channel and lower tier of ports added unto the former draught, demonstrated.

p51 There is added unto this problem the two channel wales, and also the lower tier of ports, which are in number 14, although my opinion is that 13 ports is enough on any tier for the biggest ships, and I would have been contented with the same number, only I find other nations to add more on shorter ships, which is the cause of this number here, being sure our timber and scantlings are stronger than theirs, and may better endure the sea with as great a number and as large ports as any, and so shall I say no more as to that. It remains now to show the reason why the channel wales retain that place, and also their scantlings. First, for their place. I bring them down so low in the midships as to let the upper edge of the ports be just with the lower edge of the lower channel wale, and for its shape, do either strike it with a bow or radius by which the other wales were created, and, if you please, with 3 inches more sheer than the others, and then have you the lower edge of the lower channel wale, which is the line ABC. And having done with the lower edge, I take $\frac{2}{3}$ of the depth of the lower wale, 18 inches, for the depth of this, and then will it be 12 inches deep, which I set off parallel to the line ABC, and strike the line DEF, and so is the lower channel wale completed. And having done with the lower, I come to the upper, leaving just so much distance between each wale as the lower wale was deep, which is 12 inches. Having set it off equidistant from the other, I strike the lower edge of the upper channel wale, the line GHI. Having now the lower edge, I take $\frac{5}{6}$ of the lower channel wale, which was 12 inches, and so will be 10 inches, which is set off from the lower edge of the wale parallel to the former, and strike the line KLM which completes both the channel wales. The use of them is to strengthen the ship in that place, being many of the bolts of the knees come in them, and also to put the chain bolts[5] both of the mainmast and foremast, which are for the shrouds and the whole strength of the mast, which are much more faster through a piece of 6 inches than the plank of 3 inches, which is the use of the lower one. The

---

3. *Crank.* Unstable, liable to lean over or capsize.
4. *In number 13.* The gun-ports are not actually drawn until the next page, when, for reasons which are explained in the text, they come to number 14.
5. *Chain bolts.* The bolts under the channels which provided an anchorage for the shrouds.

uppermost is to set the chainwales[1] on, to keep the shrouds from the side for the better strengthening of all the masts. All of which is to be observed, as well for the strengthening of the ship by keeping them so clear as the second tier of ports may cut as few as possible. Which is avoided by keeping them well down to the lower tier of ports, and pretty near together. Lastly, they be of great beauty to any ship, being finely shaped, so well great, straight, and proper in that place. And having now showed some reason for them, I leave it to your further experience for their use as you find cause. For the ports, I consider the depth and breadth, and to place them with what advantage possible for the most dexterity in handling the guns, and observing so much room must be kept from the second deck as to leave room for a clamp[2] for those beams of above 16 or 18 inches in the deep, and then you may find the depth of your port from the lower deck may be 29 inches to the lower edge of the port, and the depth of each port to be 32 inches, and the breadth of every port 34, which, all being well considered, is a very fit bigness for ships of this dimensions. And shall say no more of that until we come up to its proper place, and so pass unto the other work on the other side.

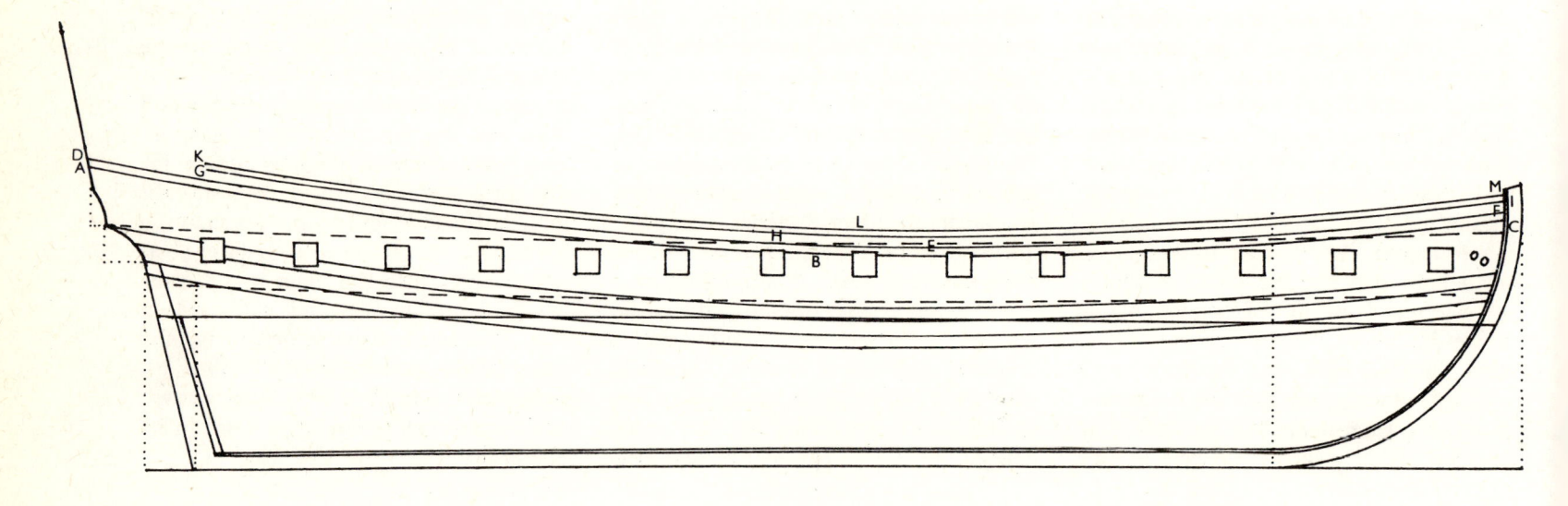

The upper tier of ports, the forecastle and quarter deck, and rising line demonstrated.

p53 There is added unto this work the ports for the upper tier and quarter deck, which are set in such order as to come just between each ports on the lower tier. The great reason for their being so placed is to place a beam just under each port, and one between, which will fall out so convenient as to place all the knees clear of the lower ports and standards where the ship is weak. Besides, it gives scarf[3] to strengthen the ship, for should all the ports be placed right over one another, it would cut the ship in sunder, which I conceive is reason sufficient for their being thus placed. And therefore I direct these ports, also on the quarter deck and forecastle, may be by the same method, so near as it may be suitable with your conveniency within board. Although I confess the exactness is not so much required but for half a foot or more for some good contrivance it may be allowed, so the greater part be placed by this order which is here set down by the ports on the whole broadside for your better direction. And I shall proceed to the drawing of the rising line both afore and abaft, being a line of great concernment as any in the ship, for by this line is the way of the ship made good or bad, it being the directing line fore and aft for the making of the ship's body fuller or leaner from the stem unto the stern, and therefore I shall take as great care to show you this line so well by arithmetic as demonstration, believing that no line unto any breadth has greater motion in the water than a true chord or circle, having myself tried both elliptical and diminishing. Although, ere I proceed I am sensible of some critics in our art will say they can make a ship's body bad or good, fuller or leaner, without the rising line which I take some care to lay down by this method. My answer to them would be, I say as they do, and then the King may have his ships want quality as these pre-

1. *Chainwales.* The original spelling of 'channel', which in this sense meant pieces of timber extending from the sides of the hull to help spread the shrouds.
2. *Clamps.* Pieces of timber inside the hull under the deck beams, which bore much of the weight of the deck above.
3. *Scarph.* Normally meant 'the overlapping of adjacent timbers in a ship's frame in order to secure continuity of strength at the joints' (*OED*), but in this case it means the placing of the gun-port so that there is no single point of weakness.

tenders do sound judgement, their ships on trial having been more ready to sail with their bottoms upward than downward, also with so much ballast as brings their gun deck into the water, or at last a good girdling, or something of greater charge, which this nation has had a good share in my time for expense. For, in short, all that can be pretended to by any, to shape the body can be but two ways, the one by rising and narrowing, the other by narrowing only. Both ways I know so well as they who can pretend to it, and therefore I shall learn the young artist a sure way which shall not fail of shaping such a body as I hope may retain no blame by the reader, and be assured by this means not to miss of my intended purpose, that the draught of 17 feet water shall not be exceeded, nor want quality to any yet built. This much I thought good to hint, to fortify the young artist against any fanatic doctrine, it being no other until they manifest theirs by views. Which as I have done these, now shall I proceed to the rising line abaft, which, being from a perpendicular let fall from C to A, which is 15 feet abaft the post on the keel, and to B on the keel 85 feet, which I make my basis, and do multiply it squarely unto itself as thus — 7225. This product 7225 I divide by the perpendicular from A to C, 17 feet, as thus, — 425. Unto this 425 I add the perpendicular 17 feet, as thus — 442, which 442 I divide by 2 thus — 221. This 221 feet is the true centre or radius of the line required, BC. And having done with that abaft, I shall also show the rising line afore by the same order, and then you may see the good reconcilement of those two lines, although by different centres, as suppose from B to E, 61 feet the basis, which multiply into itself as the former. This product 3721 divide by the perpendicular, which is 11 feet from E to F, and is derived from the rising line abaft, and is $\frac{11}{17}$, or near $\frac{2}{3}$, of that height, now being divided by 11 as thus — 338. To this quotient is added 11, the perpendicular, as thus — 349. This 349 divided by 2 and it is thus $174\frac{1}{2}$, which $174\frac{1}{2}$ feet is the true centre of that part of the circle which maketh the rising line afore, BF. Now having this perfect, I desire your remembrance, by reason you will have great occasion to make use of the same method in other lines ere you finish your work, as you will find on the other side we shall proceed to, and so shall say no more in this place, but end with the rising line.

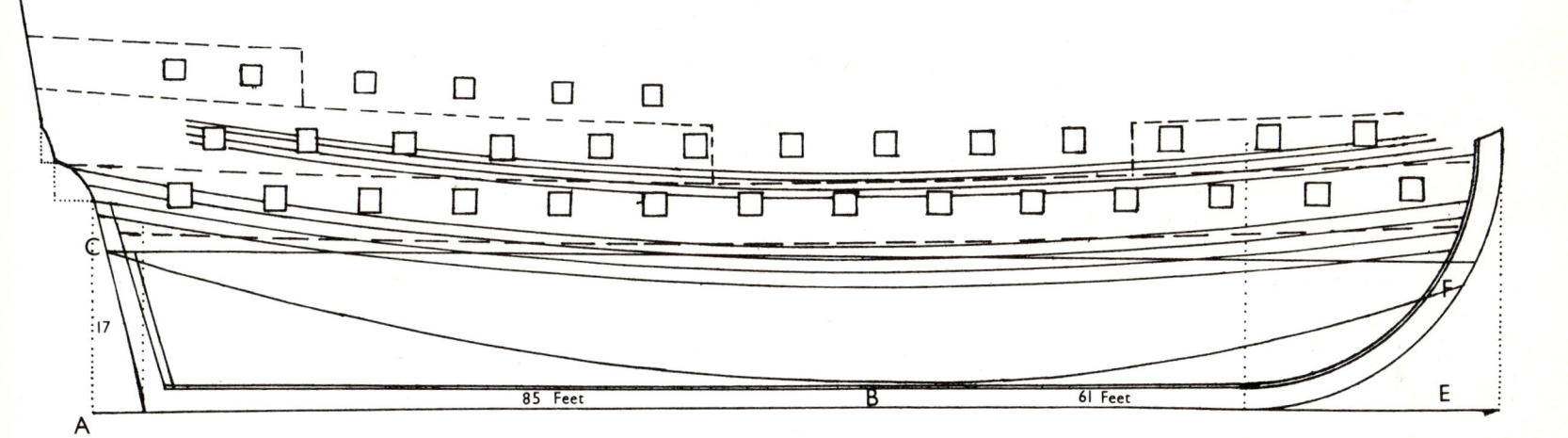

The narrowing lines of the half breadth of the floor and also the transom.

4. *Upper work ..... wrought by.* That is, the rails above the wales are either parallel to the wales, or have the same curve.

p55 In the last problem you were shown the rising line of the floor, and by these you shall have the narrowing of the floor also, together with the narrowing of the greatest breadth from the main transom to the stem, which we call the breadth line at the height of breadth. All of these lines do bear harmony one with another, as you shall see in its due order when we come to sweep out the bends of timber, until which time I must crave your patience in showing you only how to raise them, without which you cannot effect what I intend. Nor indeed are they so proper here until we had finished the plate up to the rails, and done the whole side up and down, but when I consider the multitude of lines which will be then drawn, which may hinder the distinction of the most important, I conceived it not amiss to put it well into your mind ere the rest be completed, although upper work even to the very rail, it receiving no other shape but by the same work and sheer as your lower wale was wrought by[4], which is indeed the very cause why I have demonstrated this in this plate for your clear view. And so proceed to my intended purpose. First observe that all parts of a ship are wrought from a square, and so half the breadth of this ship, being laid down, is enough, the other half being the same shape, which is needless to be shown. So that you may remember the ship 36 feet broad, the half of which being 18 feet, I set from the straight line of the bottom of the keel and bound it with the two lines which are the extent of the breadth of the ship, let fall by a perpendicular from A to B and from C to D. And having done that, I take $\frac{1}{3}$ of this half breadth, which is 6 feet, for the half breadth of my floor, and strike it as the former, being bounded also by two perpendiculars let fall from the rising line, and is EF and GH. Having thus set off the half breadth of the ship at the greatest breadth, and of the floor, and bounded them at each place, I proceed to the same method as in making the rising line, for the narrowing line of the floor, beginning at I, seeking how many feet it is to F, which I find to be 80 feet. This 80 feet I make my basis, and square it unto itself as thus — 6400. This product 6400 I divide by the perpendicular as thus — 1066. Unto this 1066 I add 6, the perpendicular C, then is 1072. This 1072 I divide by 2 as thus — 536. This 536 feet is the true centre of the chord desired, and will sweep out the line IK. By this method is the other narrowing line wrought, for forwards also, and is the line EK, which is needless to set down to increase too many figures or much writing. For before I end you shall be shown to find every part of these lines by arithmetic throughout the whole ship in their due order, but before I pass I shall show you the length of the transom, which I take from my breadth, and allow not less than $\frac{2}{3}$, nor more that $\frac{26}{36}$ for large ships, but for small I may allow $\frac{20}{27}$, by reason it lies low and not so much charge to a great ship, for breadth to the stern is a grace to any ship, and adds room, although not much quality. I do therefore reconcile all conveniences and inconveniencies so well as I can. Having set off 12 feet for the half length of the transom, it being too narrow at the line M, and work as before shown, which creates the line MN. Again, I begin at the line O and set off 10 inches for its narrowing at P, and work also by the same rule as the other. And for the bow line from P to Q I do the same, and thus have I given you the several ways of making the greatest breadth line from the transom[1] to the bow and stem, together with one example wrought of the narrowing lines abaft, all which I conceive sufficient for this plate, but methinks I hear some say that the ship would sail better if this breadth line, which I have made in several sweeps, were in one sweep,[2] it would have better motion, by my own doctrines. My answer would be, that it would be so, but here are the inconveniencies. You would have not room in the ship abaft for the cabin, nor yet in the bow, nor would she stow her guns, or bear so good a sail, and would sink deeper in the water. All these inconveniencies for a man-of-war is such that it is better to want some little matter of sailing than want room for the guns, cabins, stowage in hold, and her decks too near the water, all these being contrary to the intention of a man-of-war, which I thought to mention. And so proceed to the other matter until it be ended.

1. *Transom.* English ships were round-sterned, not flat- or transom-sterned by this date, but the horizontal beams which shaped the aftermost part of the hull were still referred to as transoms.
2. *Several sweeps ..... one sweep.* The advantage of Deane's method is that it forms a fuller hull, more able to bear a heavy armament.

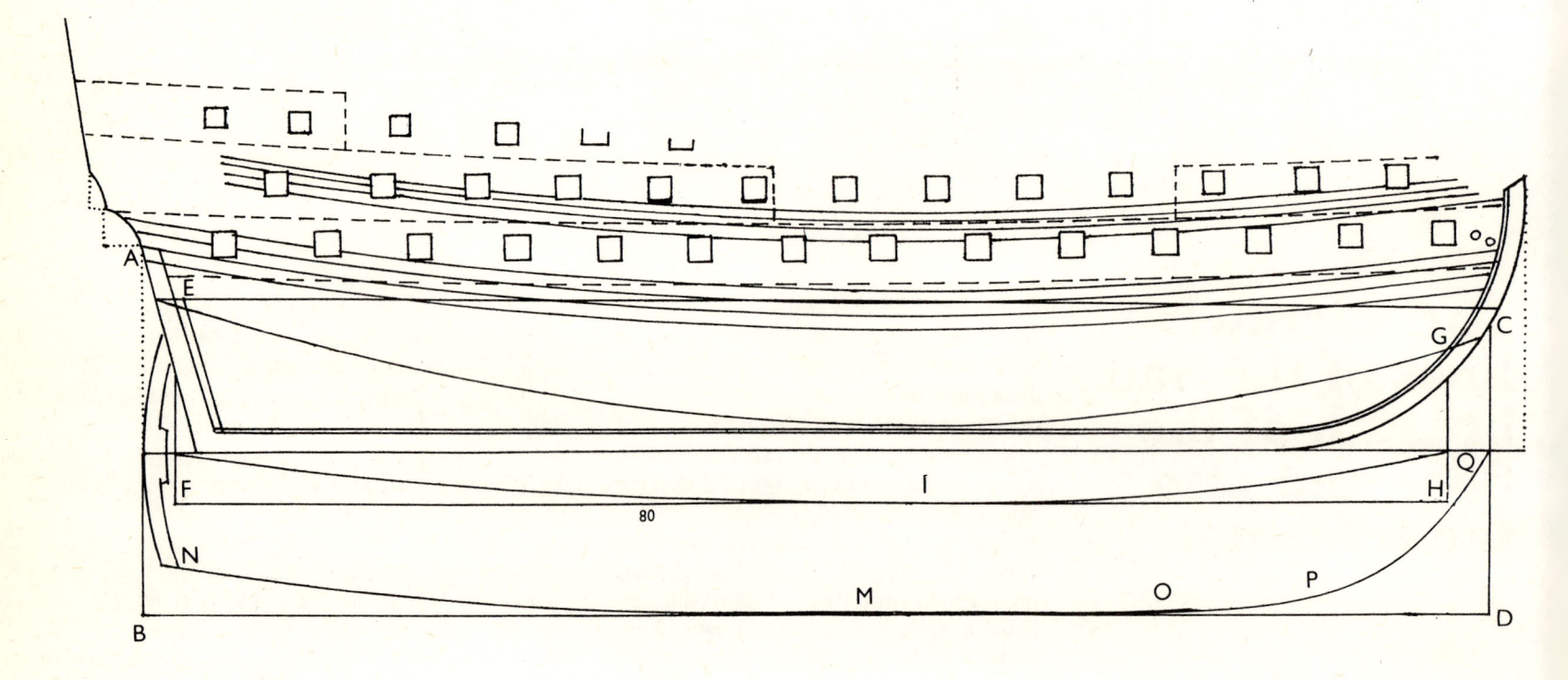

The rails of the side and head, with the knee.

59 In the last, you may remember, some little digression was made to show you the use of the rising and narrowing lines, but in this we proceed to the perfecting where we left off, and do show you the whole work from the channel wales to the top of the side, and also the shape of the knee of the head and its steeving, together with the length and shape. First for the wales, which you see from the channel wale to the top of the side. They are of several bignesses, although of no new shape, but by the same sheer of the lower wale as you were taught by the bow or arithmetic. Next to the channel wale is a great rail of about 8 inches deep, and is to be coloured yellow, above which all the black work is to be set, by which means the ship retains a great ornament, it being the greatest use of the rail AB. And for the other rails, which go all along to the top of the ship's side, and are two in number, they are for ornament, and these we call the several drifts, whereunto several pieces of carved work is added for the greater beauty. Which finishes and completes the whole upper work, being placed for depth so as to cause a sufficient height for the steerage, forecastle, coach and round-house, as you will more plainly see by the several deck lines pricked for that purpose —

The forecastle C and D and also the quarter deck line marked with EF, and the coach line marked with GH, all which being set off so high from the upper deck, with addition of the rounding of the beams, will make it 6 and a half feet from plank to plank in the place where it is proper, and more in the great cabin, and less in the coach.[3] All which have no certain rules but as the contriver sees fit, and there, finding these several heights, you end your drifts to the best advantage, with this regard to all your work, keeping it as snug as possible, remembering the depth of any ship's upper work is a great enemy to sailing, as when you have experienced this much you will confess it. So that, in short, from the channel wale to the top of the side, the decks are the greatest measure you ought to go by as to the depth of the waist and drifts, which you are still to remember are for shape of the lower wale, only something quicker[4] abaft if you please, it being fancied by many. Next you have the knee of the head, which is placed so as to let the upper cheeks some under the hawses[5] for their better strength, and withal to keep it as low as possible, to make the handsomer head by the greater hanging of all the rails[6] as you will better see by the next problem. Having given you the reasons for its place, I divine its length one half of the ship's breadth, which is 18 feet. And for its steeving I set a square stroke from the stem to the line EK, and take $\frac{2}{3}$ of the knee's length for its steeving, which is 12 feet. Which I set perpendicular from K to L. Having done my steeving and set off the length, I sweep it by two sweeps, the one of 26 feet sweep, the other 13 feet sweep, by reason the foremost end rounding the more makes a great complaint in the upper rails, as you will find hereafter, the rounding of this being from a straight line near 31 inches. All which being laid down is ground sufficient for the sweeping out the inside of the knee, which is the line NM. Now for the outside. Keep it $\frac{1}{3}$ of its length for the depth in the thwart, and $\frac{1}{9}$ of that at the end, striking such a sweep as may suit your gripe, and shape the knee as is before you, but you shall not fail to find the whole head with all its rails in its due place, with the arithmetical work complete. And so I do pass to the other work intended.

* Pages 57 and 58 are omitted in the original.
3. *More in the great cabin and less in the coach.* At that time the great cabin, on the after part of the upper deck, was used by the captain, while the coach, on the quarter deck, was used by the master and the first lieutenant. Hence the great cabin is intended to be higher.
4. *Quicker.* Steeper.
5. *Upper cheeks ..... under the hawses.* The lower rail of the head must miss the hawse holes, which are for the anchor cables, but if it comes close to them it will help to give it strength.
6. *Handsomer head by the better hanging of the rails.* The type of head which Deane describes and illustrates was already out of date by 1670 (see Introduction).

Note: The text on page 59 does not give enough information to draw the head fully, and it can only be done by reference to the diagram. The narrowing lines are omitted in the original.

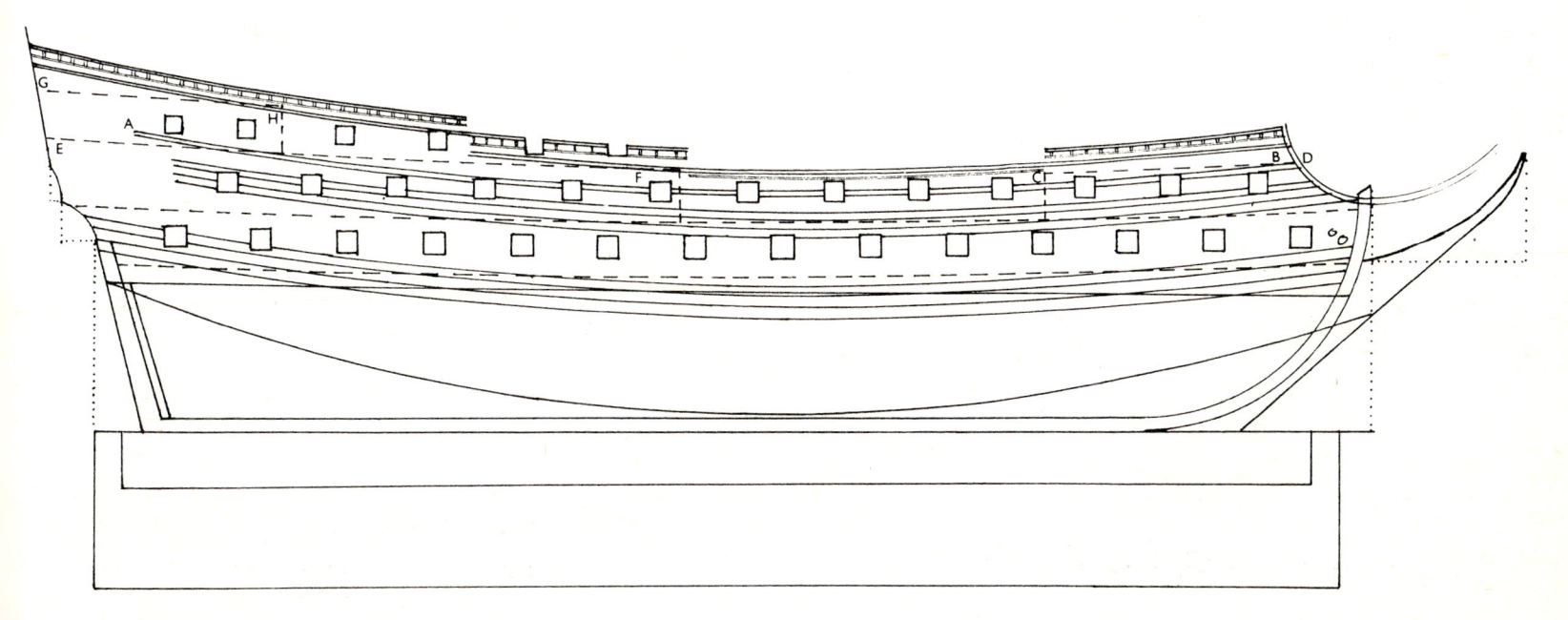

The reconciling of several sweeps demonstrated.

p61 There is nothing more advantageous in the whole work of a shipwright than the well understanding of the several lines on this page, the use of which is worth your knowledge, and therefore I have chose to lay it down here, rather than hereafter, by reason it is a work which you cannot pass by without keeping you in ignorance, which I am not willing to do, being contrary to my end. And intended this work to distinguish the several lines which you find in every page, some having the names of rising lines, some called narrowing lines, some toptimber lines, which may be shown on the plate at once, but not so obvious as you shall see here, one or two examples wrought, by which you may see all the rest more plainer. And we will suppose a line struck at 120 feet in length from A to B, where we will suppose a perpendicular is raised at C, 80 feet from A, and we will conceive this to be a keel line of that length, raising another perpendicular at A to D, 18 feet, by arithmetic. Work thus. Multiply 80 by 80, the basis — 6400, and divide this 6400 by 18, the line AD 18 feet — 355. To this 355 add 18, the line AD — 373. The half of this 373 is the centre required, which is 186 feet and a half, and if you set one leg of your compasses in E it will sweep the line FGD. Having done with that line, I would know the centre of the other part of the line from C to B, which is 40 feet or parts of a foot, and the line BH, 12 feet. I work as above, multiplying 40 by 40, the basis — 1600. This product 1600 I divide by the line BH, 12 feet, as thus — 133. Unto this 133 I add the line BH, 12 feet, as thus — 145. This 145 feet or parts of a foot I divide by 2 thus — $72\frac{1}{2}$. This $72\frac{1}{2}$ is the true radius, which, if you set one leg of your compasses in K, it will sweep the line LM, and thus have you these two centres reconciled at C, by which you may see the great harmony each part bears with the other, for had you not this knowledge your bow would deceive you, which now you correct not, only that but all other errors.

Before you cease in this work, I will show you one instance more. Supposing it an upper rail of a head, which we will suppose is 26 feet by a straight line, as you see the line AB, 26 feet or parts. On this line we let fall a perpendicular from C to D, 9 feet or parts, which is 16 feet from C to B, then I work thus. Multiply 16 by 16 as thus — 256. This 256 divide by the line CD as thus — 28. Unto this 28 I add the line CD, which is 9, as thus — 37. Half of this 37 is the centre required, $18\frac{1}{2}$, which if you set one leg of your compasses in E shall strike the line CD. Now that part is done, the other is wrought by the same method, as suppose from A to C, 10 feet, then I say 10 times 10 is 100, and divide it by the perpendicular 9 as thus — 11. Unto this 11 add 9 as thus — 20. The half of this is the centre, which if you set one leg of your compasses in K it will strike the line from D to A, which I hope you now sufficiently understand, not that I assign these lines to any use, but suppose they were of these dimensions, for if they be thus or of any other, this method shall them good in every proposition.

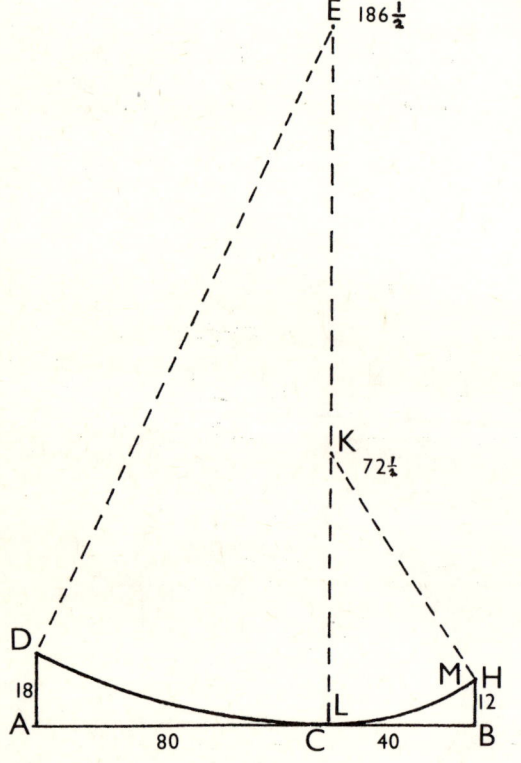

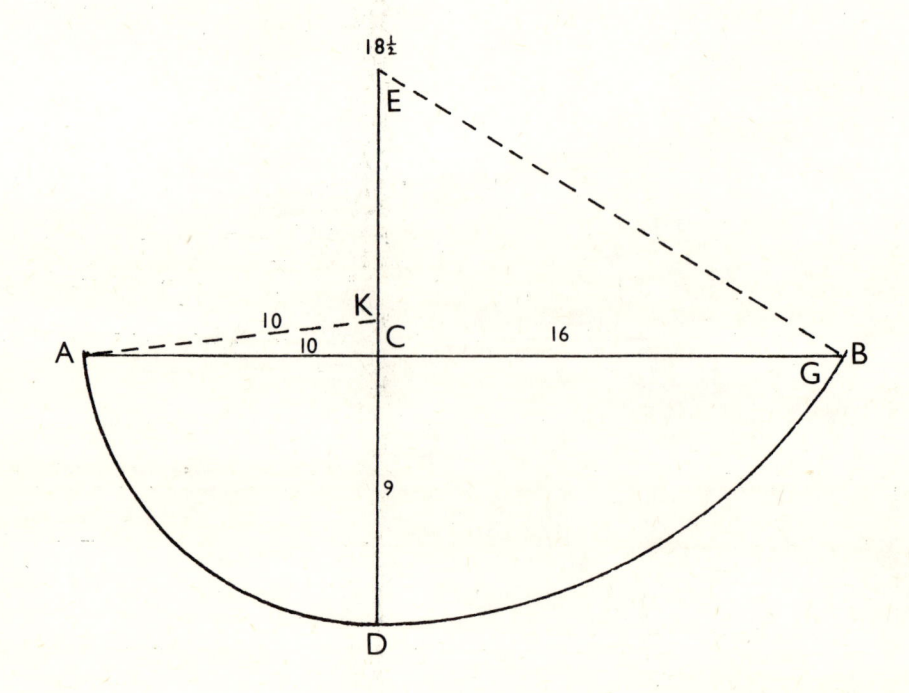

In the last plate we came from, I did entrust you in the way of working any lines by arithmetic, but as yet we have not taken any notice how to find the true rising of any timbers after the lines are found out, or the narrowing also, which I will here hint, on proposition to find out any timber's rising or narrowing without any bow or draught, which suppose to do thus. You see the line AB, which is 80 feet, which we will conceive the length of the keel from the midships to the stern post, and the line AC, the perpendicular of the post or the like, 22 feet. I then work as before taught, multiplying 80 by 80 which is 6400. This I divide by 22, the line AC. The quotient is $290\frac{20}{22}$. Unto this I add 22, the line AC, and it is $312\frac{20}{22}$. The half of this 312 is the centre required, which is 156 feet, which if you set one leg of your compasses at E it will strike the chord or line from B to C. Now you have the basis AB, 80, and the perpendicular AC, 22, and the third side BE, 156, I do demand how much the timber shall rise at K, being from B 60 feet, which I work thus, multiplying the side BE, which is 156, unto itself squarely, and it is 24336. Having so done, I multiply my side BK, 60, unto itself as the other, and it will be 3600. This lesser sum 3600 I subtract out of the greater, and there remains 20736. Having done this, I extract it by the square root, as thus — 144. This root of 144 I subtract out of the line BE, 156, and there remains 12. This 12, being set off from K to L, is the just rising of the timber which I desired should stand in that place, K. This is worth your observation, which, if you carry in mind, it worketh all the rest. Yet I will show you one example more, as suppose I would know the rising of the timber at M, which is from B 40 feet. Then I say as before. Multiply the side BE, 156, unto itself, and it yields 24336. Having done so, multiply the side MB, which is 40 feet, unto itself and it will be 1600. This 1600 subtract out of the above 24336, and there remains 22736, out of which you must extract the square root as thus — $150\frac{236}{300}$. This 150 with its fraction you subtract out of the side BE, 156, and there remains $5\frac{236}{300}$, which is the true rising from M to N. The fraction in many places is omitted to keep the memory unburdened, and thus have you two examples. The rest is the same work, and therefore I say no more, but leave it to your own industry as you find cause.

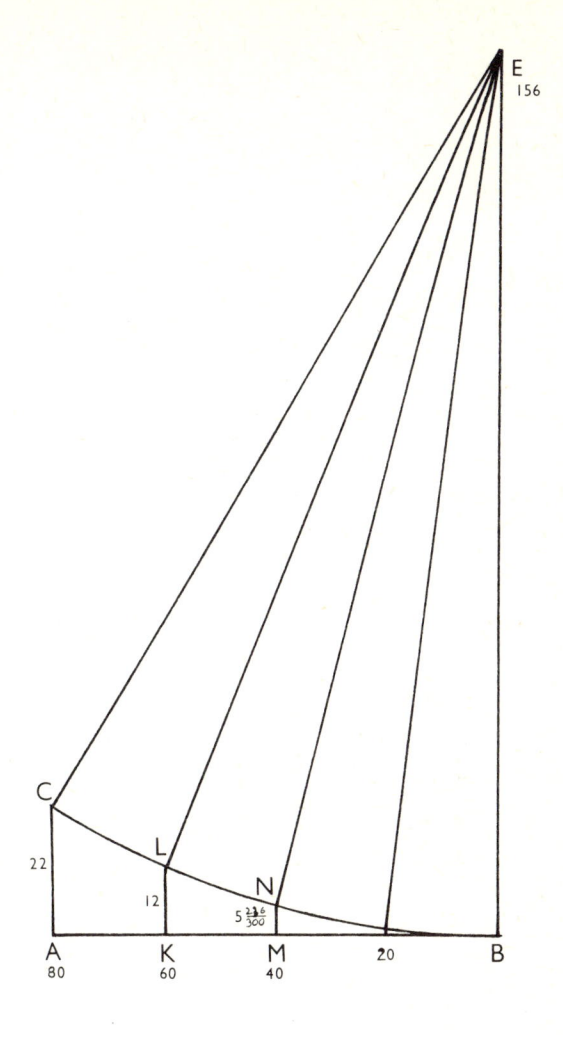

The ship's broadside finished, and lines of breadth.

The addition unto this is the rails of the head and the galleries and the carved works, which the part here, together with line called the height of breadth line, which for distinction is drawn with red and is marked ABC, being the most considerable line in this plate, the other matters being more for ornament than anything else, for which cause I shall say little as to them, they being of less use than these which I proceed to for your better instructions. And so I shall begin with the height of breadth line ABC. This line, which you see assigned at B near between the middle of the lower wales, and runs from B to C something above the lower edge of the upper wale, and from B to A just at the upper edge of the lower wale, is the greatest breadth of the ship fore and aft as you see that line drawn, and is called the height of breadth, its use bearing correspondence with the line DEFG, this latter being the greatest breadth of the ship from the several heights which the former line ABC doth bear. So that the use of them both is to set off the several bends of timber, as you will find accomplished in the next problem, the line ABC being the true height where the line DEFG doth assign the true breadth, so that the one is proper to the other, as you shall find in the next problem, where you will see more lines added together, ere you can proceed to the bends. And for its shape, it is wrought arithmetic as you have been instructed. Next you may see several perpendiculars which are marked with letters and figures, they being the just distances or spaces of every third timber to stand on the keel when they are perfected, and are called the bends of timbers, which being once swept out are set on those several places of the keel on which they stand, making the shape of the ship in that place, and they are called frame bends, which is proper to their being the guidance of the whole body, so that I shall say no more here. Having told you the name and use of each line, which I hope you will remember when you come to further view of the others, all which lines, both perpendicular lines, height of breadth line and the narrowing of the breadth line, with several others, must be handled at once, for which cause you must have them here singly for your clearer understanding, before too many lines are multiplied to spoil your view, or create you trouble, which is all at present as to that. We now come to finish the head, you having been in the last shown only the knee. Being once set, and a lion at the end thereof, about 13 feet in length and about $2\frac{1}{2}$ feet in depth, proceed the sweeping of the upper rail as thus. You having considered the depth, how much the rail shall be from the knee to the upper edge of the upper rail just at the stem, which is $\frac{1}{3}$ of the knee's length, which will be 6 feet, you will then find your hanging of the rail from a straight line to be about $8\frac{1}{2}$ feet, which greatest hanging must be near the outside of the stem, as you find. And having thus set off your three assigned places at NMO, work thus, as you have been showed, which I once more set forth for your better memory. Finding from P to M 8 feet and a half, multiply it unto itself thus — 72. This 72 divide by 8 feet from N to P, the quotient will be thus — 9. Unto this 9 add the $8\frac{1}{2}$ the perpendicular, and it is 17, which divide by 2 and it is the radius required, which is $8\frac{8}{14}$, which is the centre of the after part of the rail. The former is like unto this. Supposing from P to O, 12 feet, which is the basis, multiply it squarely — 144. This product divide by the perpendicular from M to O, the greatest hanging, as thus — 18. To this 18 add the perpendicular $8\frac{1}{2}$ as thus — $26\frac{1}{2}$. The half of this is your radius required, which is $13\frac{1}{2}$, where you may see the great exactness and good reconcilement at once in these two sweeps for the rail, and the truth of this work, which I hope by this you fully discern. The other rails are divided and are wrought as the upper one. Thus have you all that is added unto this work of any concernment. The rest is the beautifying, which is left to pleasure, what cost shall be bestowed. And proceed to the next, where you will find the whole ship finished complete.

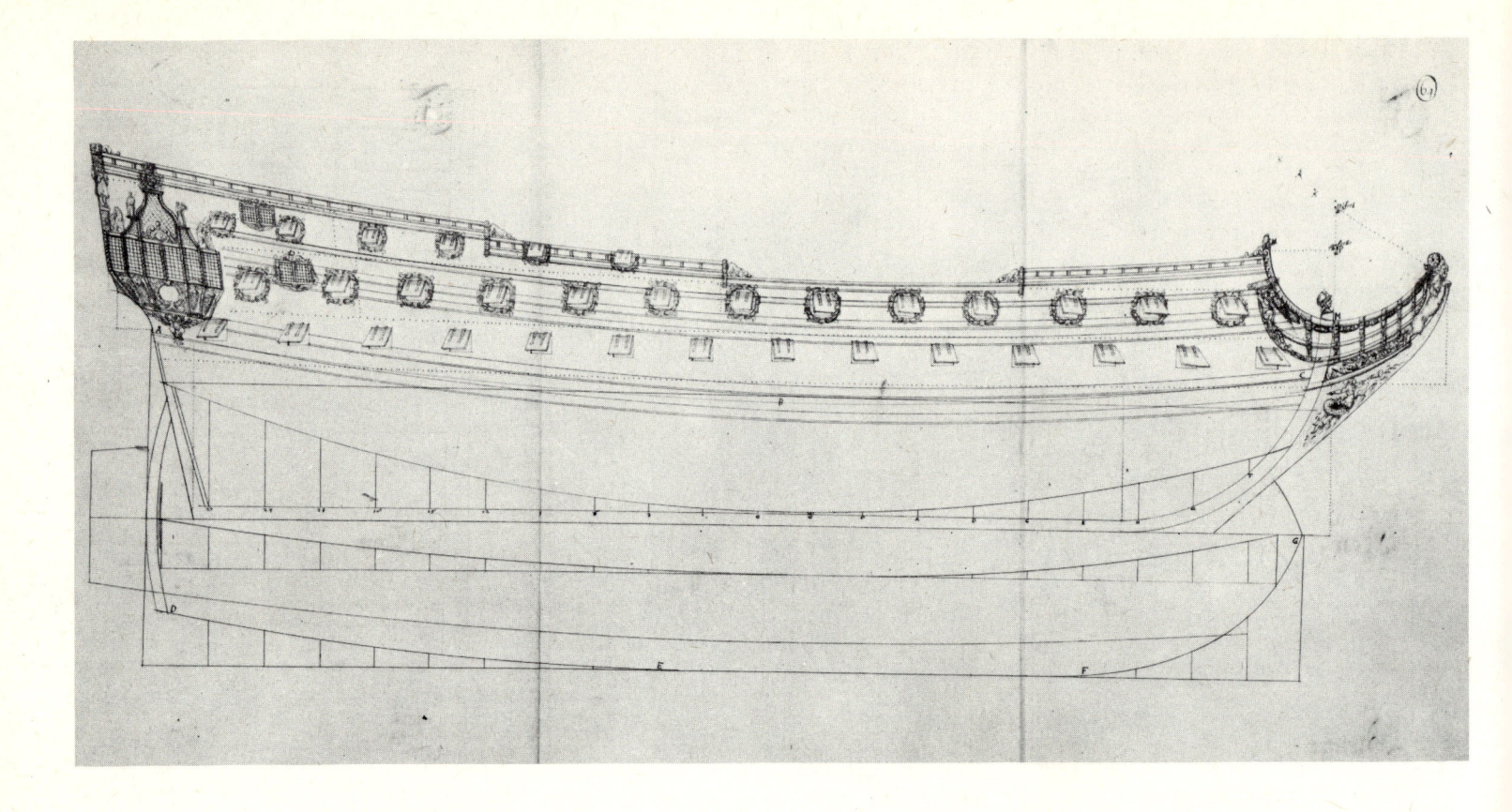

Note: This is the only drawing in the work to show round gun-port decorations, which were far more common than square ones by this time.

To draw a bend of timber in the midships.

p65 In the last page you were instructed in the several narrowing lines, and also the height of breadth lines which relate unto the several bends of timber, which you will find perfected in the following problem, where you will see the whole draught, with every line and bends of timber drawn as they ought. And fearing the multitude of lines may hinder your understanding and cause it to seem dubious, I shall therefore in this whole sheet show you the way of beginning how to draw the bends of timbers, which when you have seen will prepare you to apprehend them, where they be drawn one within another, and how they be wrought. As, for example, suppose I would strike out the first bend of timbers in the greatest breadth in the midships, which I do thus. Take from your draught the half breadth of the ship, which is 18 feet, and strike it on the place you intend, as thus, in the line AB. This line is derived from the narrowing line of the greatest breadth, which in your last problem is the line DEFG, 18 feet, as you have often heard.

Now having struck this line, the half breadth of the ship, mark it with AB. I make that line my basis for the rest, and do raise a perpendicular at the end of the line at B to the line C, as thus. Which perpendicular shall be the middle line to guide the whole frame, setting off all other breadths from that line.

Having done the perpendicular BC, I take from my draught the just half breadth, which is 18 feet. Setting one leg of the compasses from C to D, and from B to A, I strike the line DA, parallel to the line CB, always remembering these two perpendiculars be so deep as the ship from the upper edge of the keel to the top of the side. And thus is the half breadth of the ship bounded by these two perpendiculars, BC and AD.

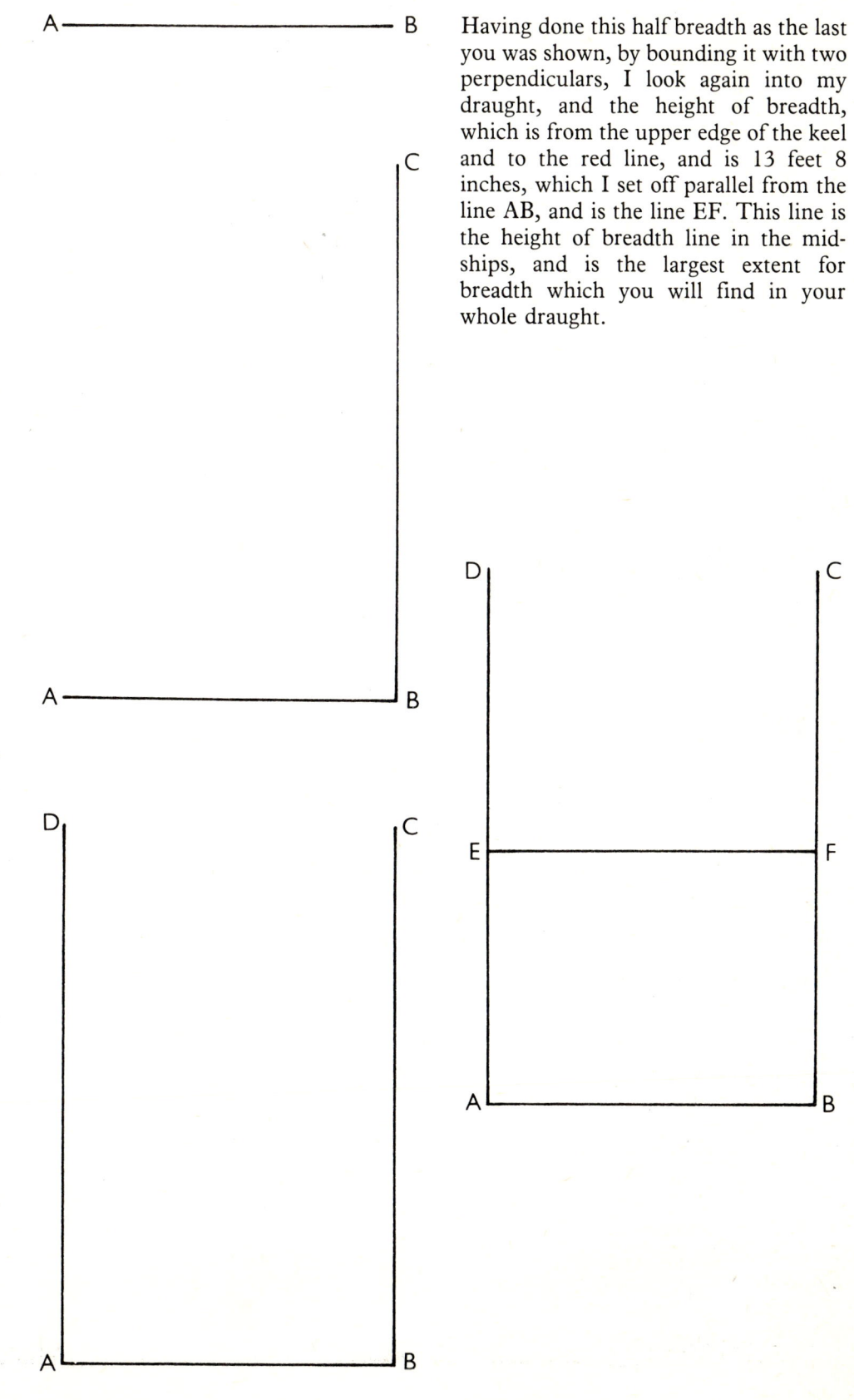

Having done this half breadth as the last you was shown, by bounding it with two perpendiculars, I look again into my draught, and the height of breadth, which is from the upper edge of the keel and to the red line, and is 13 feet 8 inches, which I set off parallel from the line AB, and is the line EF. This line is the height of breadth line in the midships, and is the largest extent for breadth which you will find in your whole draught.

Having done with the former, you look again into your draught and take the half of the floor, which you have been told is near 6 feet. When you have done, you set one leg of your compasses in B to H, and from F to G, and strike the line GH parallel to FB. This line GH is the utmost bounds of the half breadth of your floor, on which you will find we set the centre of the sweeps for the lower futtocks and floor timber heads, as you will soon perceive, and also bounds the floor in its utmost breadth.

Having prepared these lines, you look again on your draught, and find how much you are to tumble home your toptimber, which line you find it to be near $\frac{4}{6}$ of your half breadth of floor, which is 4 feet. Which you take off with your compasses and set from D to I, always remembering this tumbling home is at the top of the ship's side. And now have you prepared your work, to sweep out the midship bends, and have raised all your lines of the greatest extent, within which you are to set off all your narrower bends as you proceed aft, always remembering this is but half the ship. The other is like, only omitted for brevity, and is needless to set down. The like lines you must make for to work your forward timbers, they being of another shape and not like those aft. Having thus taken off these lines from your plate, I shall show you the sweeps which shall make a good midship bend.

p66 Now all the lines are prepared for sweeping out the midship bend, I take from my scale one fourth part of my whole breadth of my ship, which is 9 feet, and set one leg of my compasses in the floor line at K and sweep it from H to L. This sweep is called the floor sweep. Having done with that, I take $\frac{7}{9}$ of the floor sweep and strike it under the breadth line, downwards from E to N, and the centre M.

Having prepared those two sweeps above mentioned, I take $\frac{20}{36}$ of my breadth, setting one leg of my compasses in O, and strike the line from L to N. The sweep is the sweep which maketh the upper and lower futtock moulds, as you will perceive ere you have done.

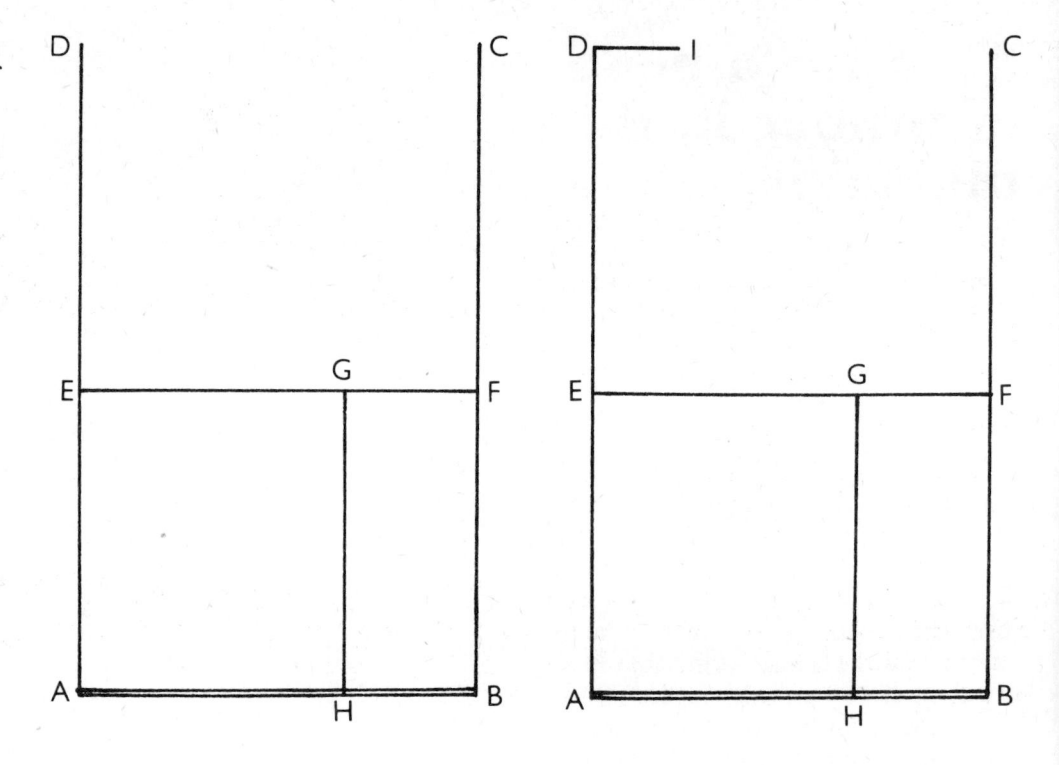

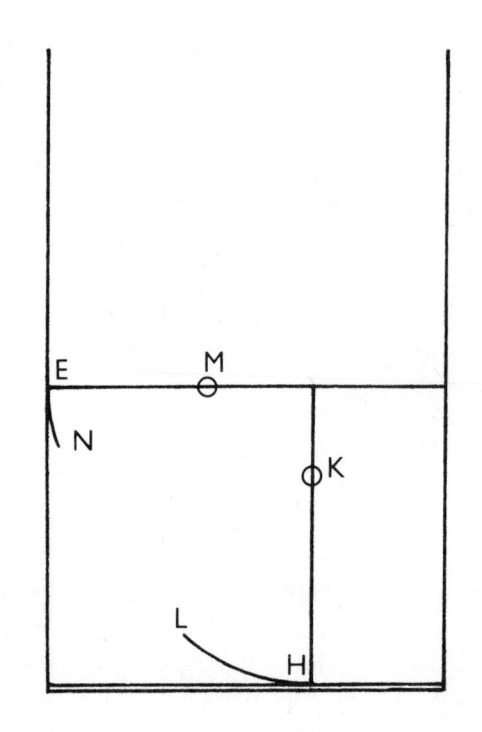

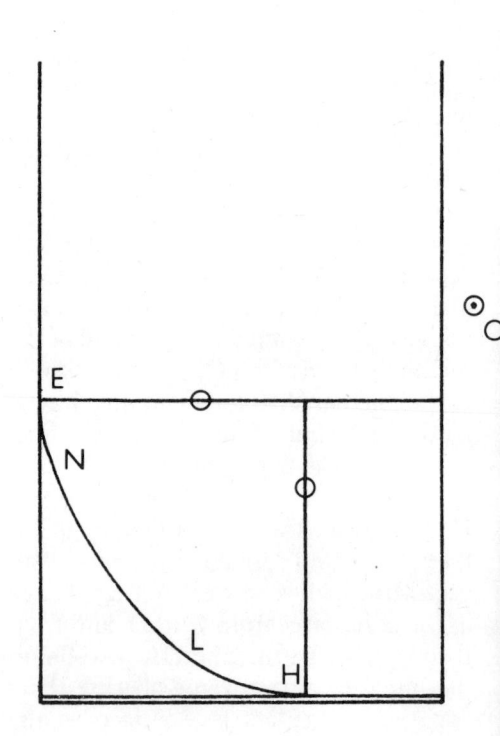

Note: The text does not mention the line slightly above and parallel to AB. This represents the dead rising, which is the amount the floor timber rises at this point. It can be seen on the profile (page 54) from the fact that the rising line does not touch the top of the keel in midships.

68

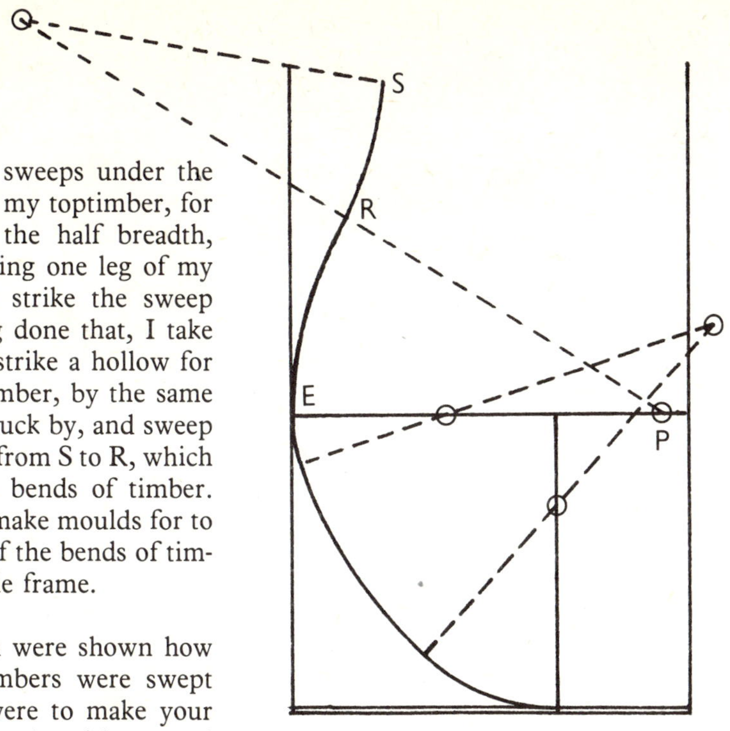

Having done all my sweeps under the breadth, I come unto my toptimber, for which I take $\frac{17}{18}$ of the half breadth, which is 17 feet, setting one leg of my compasses in P and strike the sweep from E to R. Having done that, I take the same sweep and strike a hollow for the head of the toptimber, by the same radius the last was struck by, and sweep out the hollow sweep from S to R, which sweep completes the bends of timber. By which you are to make moulds for to graduate all the rest of the bends of timbers and for the whole frame.

On the last plate you were shown how the half bends of timbers were swept out, by which you were to make your moulds, and being made with a good scarf for every timber. You are to proceed in the sweeping out the remaining part of the ship aft, as you find one within another where the ship is completed. But for fear raising them may be dark to the understanding, I will show you one example more to make you perfect. And shall suppose I were to raise the bend of timbers as 15 which I do thus. I look at my draught and take the rising from the keel to where the line 15 stands on the draught, which I set off from the line AB, and is the line CD.

Having done that, I also take from my draught the narrowing of my floor, which I set off the line DE, and is the line FG. This is the narrowing of floor.

Having done that, I take from the draught the height of breadth at 15, which I set off from the line AB, and is the line HI. Having done that, I take from my draught the narrowing or greatest breadth at 15, which I set off from I to H, and the like narrowing and rising for the toptimber head, which is the line KL. Now having set off all the narrowing and heights in every place, I proceed to sweep the floor sweep 9 feet as the former[1], setting one leg of my compasses in P and sweeping under breadth from Q to R. Having done that, I take the same 20 feet sweep of my midship bend and set one leg in S and sweep from R to O. Having done that, I keep my centre for all the toptimbers in the midships, setting one leg at T and sweep from Q to W, and from that the hollow sweep as in the other, observing to fetch out the hollow at the stern. All the sweeps being thus struck, you have the half bends of timbers completed at 15. In like manner are all the other.

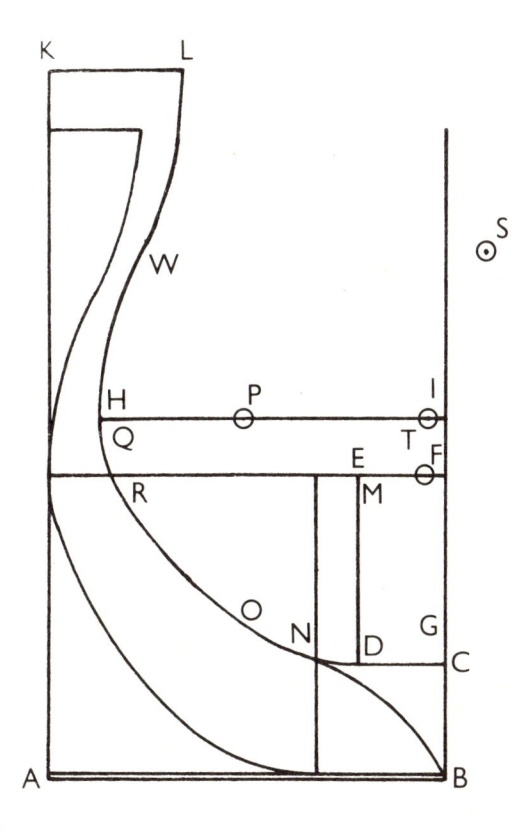

This latter is no other than the former for its nature of working, only as the last was the bend of timbers aftward on, marked 15, this shall be a bend of timbers forward named N, which is set by the narrowings and risings as the other, only as the one is worked aft on the starboard side, the other is worked aft on the larboard side, that one suit of moulds may serve your turn to build by. As for example I look on my draught and take the true rising from N where it stands on the keel to the rising line, and set it off from the line AB, which is the line CD. This is the rising line. Then I take from my draught the narrowing and set from the line AE, which is the line FG. The rising or narrowing or the breadth is the same, as is often shown. Having thus done, I proceed to sweep by the same sweeps as above, sweeping out from H to I, the floor sweep, and from K to L under the breadth, from K to M above the breadth, and from M to N the top of the side, which I hope by this you see perfect the rising of the whole ship's body in every part. Only as you have these singly, the other be one within another, as you will find in the next place you come at, as appear to your better satisfaction.

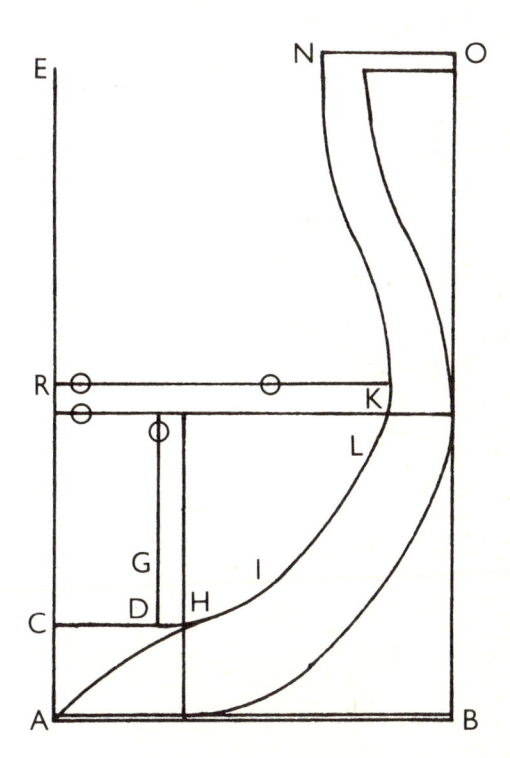

1. *9 feet as the former*. It is clear from the diagram that the radius is less than 9 feet. Most of the other sweeps are less than the diameter in the midships frame, except for the toptimbers, which are the same (see Introduction).

69

The ship's draught completed in every part.

p67 Here you find the whole body finished, with every line in its place, from which is raised the several bends of timbers, being every third timber from stem to stern, and do stand just on the keel, where you see the several letters and figures marked. They are derived from six lines, namely the narrowing of breadth, height of breadth, narrowing of floor, rising of floor, narrowing of the toptimbers, rising of the toptimbers, which is the top of the side. From all which every bend of timber receiving his part, being guided by a square as all other works are truly performed, and after they be raised by these lines for breadth and height, they are swept out by several sweeps, to make the best bends and swiftest motion, and so full or lean as to cause the ship to carry all her weight which is proper for her force. In fine, these bends of timber are the greatest art which belong to the master builder, and have so much of variety in them as you can have imagination in your thoughts, as you will find when you perceive the work. And since the variety is so great on which depends the good quality of the ship, you shall be shown a way to find every part of breadth, depth, and sweep which shall so complete your ship's bends, so as to reconcile the way of the ship to the rudder, and also a fair entrance into the water forward unto the bilge, and yet neither fuller or leaner, in no part underwater than such as to cause her to carry her proportion guns and stores, and not to exceed the draught of water spoken of, nor want quality to any yet built. In short, when you are fully sensible of this method, you shall find the body so clean underwater, and her shapes so fine above water, as nothing more can be added. And if so, it must be concluded a good essay to make a perfect builder. And because this work, now it is added together, hath many lines, I shall only name you these which the several bends are raised by, and so stands one within another, those with the figures being the half breadth aft, and those with the letters the half breadth of the several bends forward. The raising of each bend is showed you

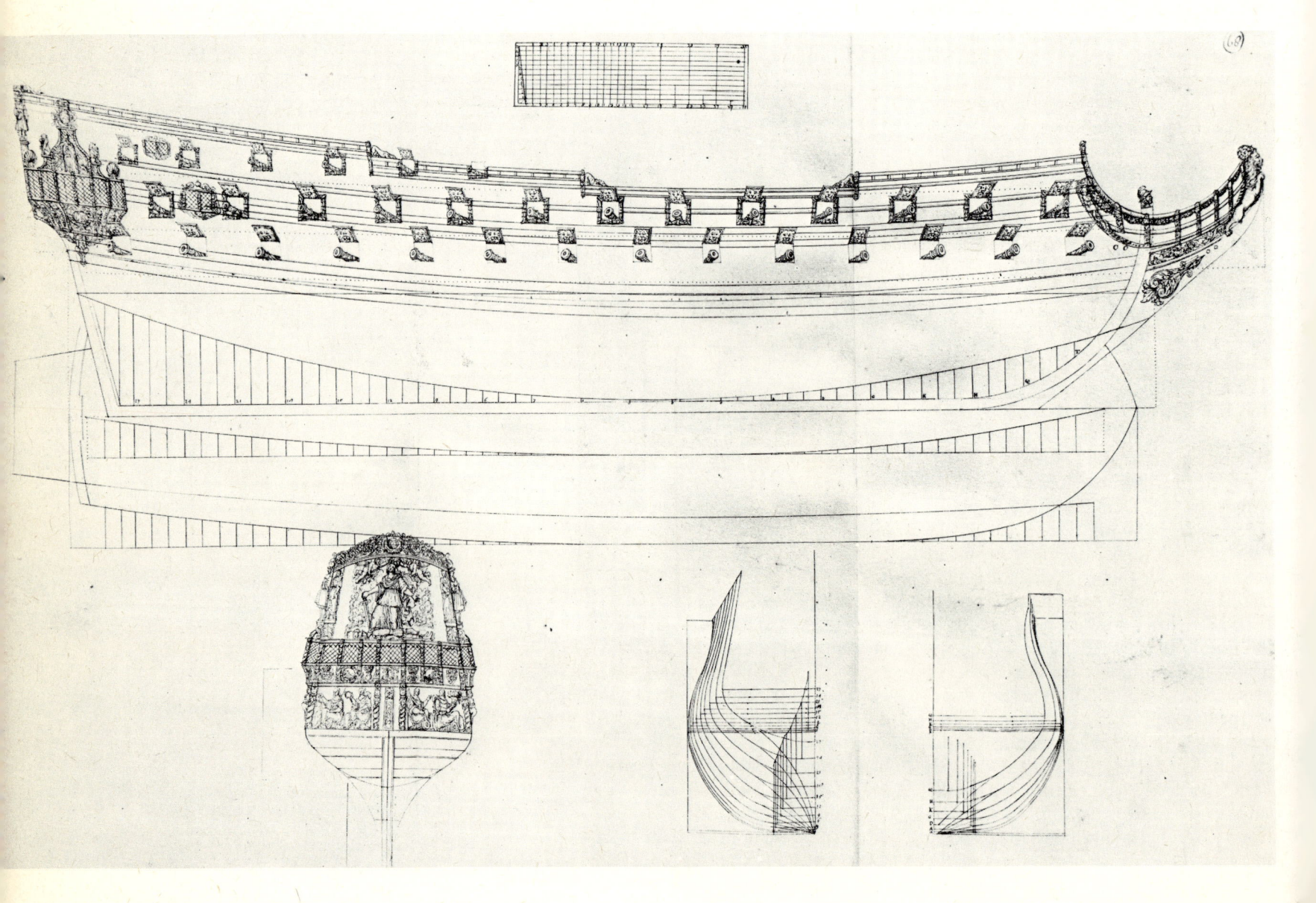

HULL DESIGN

more clear in the last leaf you came from, so that my aim must be here to show you them altogether, and let you see the body in its true shape. Which would be done could you but remove every one of those several bends of timbers and set them on the keel according to their place, as the several bends aft which is 3,6,9,12,15,18,21,24,27, and where set thwart on the keel at the same figures of 3,6,9,12 and the like, it would show you the shape of all the ship aft, which you now see only the flat. Also those several bends you find marked with A,D,G,K,N,Q, being removed from the flat to their places on the keel where the like letters are of A,D,K,N,Q, it would make the whole body complete forward on, and then you would see the figure of the ship as they are a-building, which now appears flat. And having now made and finished all the bends, put them to proof whether they be good or bad, by setting so many parallel lines beginning at the draught of 17 feet water, which is done thus. Draw a line through all the bends of timbers at the height as the pricked line crossing all the bends at LM, which done take off the breadth at every bend by his name and carry it to the half breadth against each figure where it doth belong, setting one leg of the compasses in the line of the bottom of the keel and the other where that breadth will fall. Having thus done to every bend throughout the whole ship, you will find what line these will make. This line I call the water-line, or seat of the ship at her greatest depth of water. Having done with that, I proceed to set off one more at 15 feet water, at lines NO, and also downwards until I have seen how the body is shaped, until I come near the keel. Having taken the pains, and finding all those lines go fair, making neither swell nor cling, I conclude my work to be good aft, agreeing with the old lesson where we began, every line being part of a circle and is reconciled to every breadth of the ship, which is as great a truth as can be afforded, for so much stern, bow, and the like. And until you find some better way I advise this to be practised, by reason this being thus performed shall not miss one of the qualities above mentioned, you shall still see more clearer ere I leave you. And so proceed to the next plate, to show you the reason of swimming or sinking by the hydrostatical rule, it being worthy of your pains, and indeed no-one can be a good shipwright who is not perfect in this matter, it being the dependency of our draught of water, which you shall clearly discover when you have perused it.

To measure a ship's body underwater, and the reason of the ship's swimming and sinking.

p69 The principal intention of these four figures shall be to inform you how to measure the superficial content of any ship's body underwater at any draught you intend, and withal to inform you of the reasons of swimming or sinking, which being once made plain to your understanding nothing can please you better, especially when you by this rule see all your labour perfected without error. And by this understanding you can tell exactly what weight your ship's body will weigh of itself, and also what weight shall be put into her to sink her down so deep and no more, so that you shall never fail of ordering beforehand what ballast, provisions, stores, guns, weight of men and the like shall sink her so deep, and into the water as you assign her, which I conceive you will think a good work if it be true. And therefore I shall proceed to show you the reasons by example. First observe that the water is a juster balance than any made by art, for in any balance which is heaviest, the lighter weight riseth and the heaviest weigheth down. Just so it is with all sorts of water, having in themselves such a standard of weight, as anything put into it which is lighter than the water it swims, if heavier it sinks. This standard in salt water is just $64\frac{1}{4}$ of an ounce to 12 inches square, every way of measure. In the River of Thames a foot square weigheth 63 lbs 12 oz, spring water 62 lbs $11\frac{1}{8}$, rain water 62 lbs $1\frac{1}{2}$. These latter being lighter than salt water is the cause a ship draws more water in the fresh water than salt, by so much as the water differs in weight. Now have you the true standard both for weight and measure, you must next hear what I pray. Remember in these few words that if you make one piece of green oak of just 12 inches square every way, which is so full of sap that it weigh 65 lbs, it will sink, and $15\frac{3}{4}$ of an ounce will force it up to the face of the water and there remain. Again, if you make another piece of dry oak of just a foot cube each way, and through its drith[1] it weigh but 56 lbs, then it will swim, and just 3 lbs $\frac{1}{4}$ oz will sink it to the surface of the water and no lower, and the reason is what you

1. *Drith.* Dryness.

DEANE'S DOCTRINE OF NAVAL ARCHITECTURE, 1670

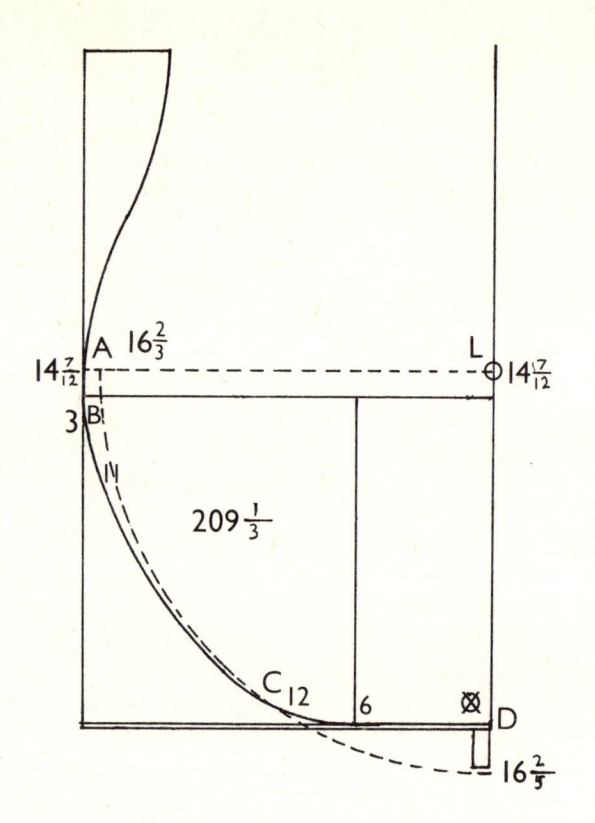

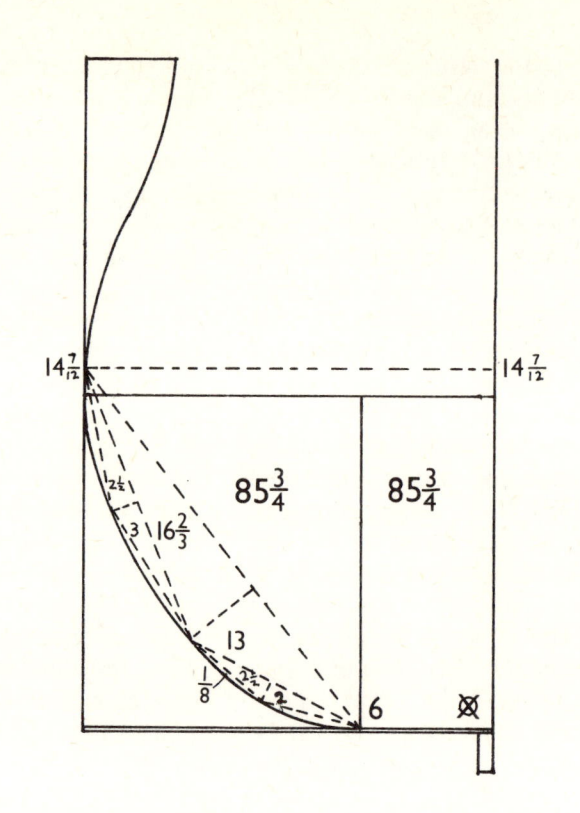

have been told. So that if anything be put into the water which occupies 12 inches cube and weighs more than 64 lbs it sinks. If less it swims deeper or lighter as it is near that weight, as myself hath often experienced, and could say more, but shall leave all, concluding with this, that if this be true, then the principle is to measure the ship's body underwater and find how many feet it contains, and her weight of body is so many 64 lbs as there shall be feet square in that body under water and no more. And as you put weight the ship increase measure just equal to what you put in, so that if you put in 20 tons it is just so many foot square as contains the measure equal to the weight, so that the figure of water which she opens, were it weighed in one scale and the ship and all things in her in the other scale, they be of just weight to an ounce. It remains now to show you how to measure the bends, which is done thus[1]. Measure from A to B, 3 feet, from B to C 11 feet, from C to D 12 feet. Add them all together, as thus — 26 feet. This 26 multiply by 4 as thus — 104. This 104 multiply by 7 as thus — 728. Unto this 728 add one cipher as thus — 7280. Divide by 225 as thus — $32\frac{180}{225}$. The half of this is the centre, which is $16\frac{5}{7}$, and will strike out a circle equal to a square with this bend of timbers. Setting one log of your compasses in L will sweep the line which is marked $16\frac{2}{3}$. Now you have the whole diameter $32\frac{180}{225}$. Multiply it by 11 and divide it by 14 and you will have 36 feet in the whole circle whereof this is but $\frac{1}{4}$ part, which by dividing the 36 feet by 4 is the content, 209 feet, contained in this midship bend when the ship is at 17 feet water from the bottom of the keel. Having done that, I show you another way by measuring the whole body, by the doctrine of angles, setting each measure in the several angles as you see in the line AB, wherein I find $210\frac{3}{4}$ feet, which agree with the former within one foot, and thus have you the work perfected, which being well observed will make you a good shipwright. I do not stand to tell you that the reason of this arithmetic lest it should take away the view of these which I chose earlier to tell you, and leave you to authors for the rest as you please to take the pains, my intention being only to show a shipwright's work and not a master of learning of which I never understood so much, as you perceive, to write good English.

1. Deane's method for calculating the displacement was of course a considerable advance for the times, but by modern standards it is rather crude. The first method (lines 24 to 44) uses an approximate method for calculating the average radius of the underwater body. Several distances along the circumference of the frame are added together, and this is taken as a quarter of the circumference of a single circle. The radius is found using the formula $2\pi r$, and the figure of 70/225 is used for π. The other method, by dividing the area into a number of triangles, was probably more accurate if enough triangles were drawn, and it was common in later times. Neither method appears to take into account the thickness of the plank.

p70 You had on the other side the bend measured by a circle equal to a square of like superficial content as it is to 14, but in this you shall have an example expressing the dimensions of each angle contained underwater in the bend whose name in the draught is 12, as you will find by its form. Supposing the bend divided into the most regular angles to the end it may be truer measured, supposing that part which is a square ABCD being 11 feet one way and $6\frac{1}{2}$ the other, then we do multiply one by the other — 71. This 71 is the feet contained in that piece as you find by the content set therein. Again, I suppose the angle EFG to be 10 feet from E to F and 11 feet from F to G, which I multiply the one into the other — 110. The half of this 110 is the content of that angle, which is 55 feet as is described. Again, I find the small angle HI to be $6\frac{1}{2}$ feet one way, and $3\frac{1}{2}$ the other, which I multiply the one into the other — $22\frac{1}{2}$. The half of this $22\frac{1}{2}$ is the content of this angle, which is $11\frac{1}{4}$. The remaining part I reduce into so many angles till at last I have no segments, which is found $12\frac{1}{2}$ feet also, which omitted doing in this diagram to multiply line and letters to spoil your view of this, which I was willing to show you, and so do not question but to be understood in what I have laid down in the other. This bend being done, the other bend which is the place of 24 in my ship is done by the same rule, and hath contained in it $47\frac{1}{2}$ in one piece, 7 and $6\frac{1}{3}$ in the two others, both which figures being thus measured and there parts added together, the one contains $150\frac{1}{2}$ feet square, the other $61\frac{3}{4}$. Having thus measured every third bend of timbers throughout the whole ship, I set them by themselves. Having so done, I must measure the length of the keel from bend to bend by his name, as from the 6th bend to the 12th may be 14 feet. Why then, if the bend 6 and 12 were of equal content it were no more but to multiply the length by the content and then that part of the ship were done, and proceed to the rest. But this would be false by reason of the ship's side as you move fore and aft is a part of a circle. You must imagine thus. Supposing that you would measure from 12 to 20, which may be 28 feet from one to the other on the keel, between which the side hath part of a circle. Then I say take two thirds of the bend 12 which is 150 feet and that makes 100, and take one third from the bend 24, which is $61\frac{1}{3}$, and it is $20\frac{1}{3}$, which add to the 100 which was two thirds of the other bend, and that shall be a mean, as thus — 120. By this mean you multiply your length of the keel and it is the true content of that part of the ship between 12 and 24. But this I only set for example, not truth to measure so far at once, but to measure every bend from stem to stern by this method and you will find all your labour true so well as pleasant. Having once summed up all the whole body underwater into feet it is just equal to so many 64 lbs in salt water, in fresh 63 lbs 12 oz as you have been taught. And thus by setting your ship in the water as soon as she is launched she will draw 12 feet 6 inches abaft, and 8 feet 8 inches afore without one stone of ballast, mast or rigging. Having discovered this you can easily know how much you must put in by measuring the ship's body from that water she drew light to your assigned depth of 17 feet water as you have been shown. I could tell you why this ship draws no more light, but in another place it shall be done to prevent much trouble here, where I would have taken good notice of this work.

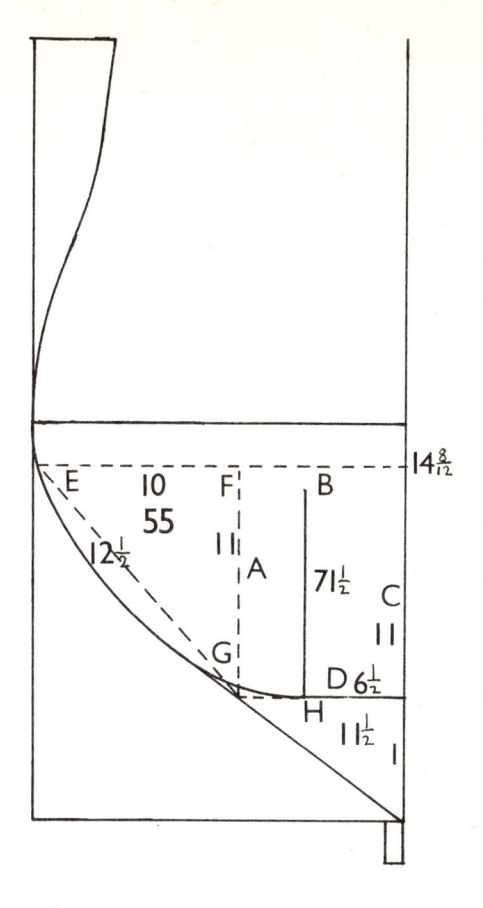

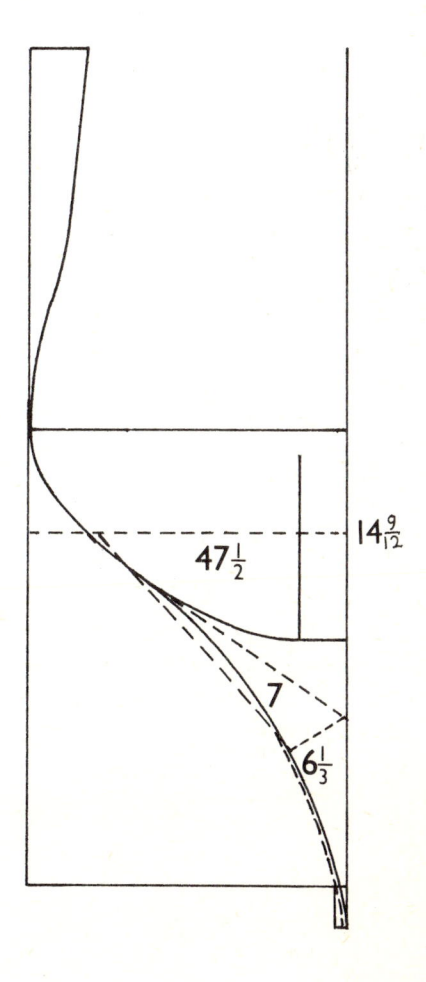

Every place and piece in the ship alphabetically demonstrated.

p71 This page shows you all the several pieces and their names throughout the ship, so far as they can be seen on the plate. Yet there are some that cannot be discerned without a particular description, which are not many, yet, so few as they be, I will not pass them by without your knowledge. The first is the foot-waling, that is, the several planks which is the sealing[1] of the ship in hold, brought over all the timbers within board as the plank is without board. Of these there be three sorts. Them on the rungheads[2] some call sleepers, others footwaling, those between the deck and the rungheads, middle-bands. Those the beam ends rests upon are called clamps. There is nothing more in hold but what you will find in the abstract, except futtock riders, which for shape is as the futtock timbers[3], only they are bigger and are brought over the footwaling as the floor riders are. The remaining parts that cannot be shown are standards, up and down knees which fasten the ship's side and the beams together, and thus have you the most considerable names and pieces about the whole ship.

A	The Stern Post
B	The Keel
C	The Stem
D	The Gripe
E	The Knee of the Head
F	The Lower Wale
G	The Upper Wale
H	The Lower Channel Wale
I	The Upper Channel Wale
K	The Great Rail
L	The Top of the Side or Gunwales
M	The Upper Rail of the Head
N	The Lower Counter
O	The Second Counter
P	The Upright of the Stern
Q	The Mizzen Mast
R	The Mainmast
S	The Foremast
T	The Bowsprit

1. *Sealing.* The planks which lined the hold. The spelling used by Deane may give a clue to the origin of the term. In later years it was spelled 'ceiling'.
2. *Rungheads.* The ends of the floor timbers, which were the lowest parts of the frame timbers. Probably they were so called because after they had been fitted the structure resembled a ladder.
3. *Futtock timbers.* The upper parts of the frame timbers.

HULL DESIGN

1 The Kelson	20 The Bulkhead of the Great Cabin	42 The Standards
2 The Step of the Mizzen Mast	21 The Steerage Bulkhead	43 The Up and Down Knees
3 The Step of the Mainmast	22 The Jeer Bitts	44 The Cheek of the Head
4 The Step of the Foremast	23 The Topsail Sheet Bitts	45 The Trail Board
5 The Floor Riders Ends	24 The Jeer Capstan	46 The Hawses for the Cables
6 The Pillars for the Main Capstan Step	25 The Bulkhead of the Forecastle	47 The Figure of the Head's Place
7 The Shot Lockers of the Well	26 The Beam Ends of the Forecastle	48 The Supporters of the Catheads
8 The Ends of the Orlop Beams	27 The Jeer Bitts	49 The Steeving of the Head
9 The Ends of the Gun Deck Beams	28 The Topsail Sheet Bitts	50 The Kelson of the Head
10 The Gun Deck Ports	29 The Knees of the Bitt Pins	
11 [No figure visible]	30 The Catheads	
12 The Main Capstan	31 The Coach Bulkhead	
13 The Ladder to go on the Upper Deck	32 The Beam Ends of the Coach and Roundhouse	
14 The Jeer Capstan	33 The Main Beams of the Gun Deck	
15 The Main Bitts	34 Are All the Small Ledges	
16 The Standard against the Bitts	35 The Carlines of the Gun Deck	
17 The Ends of the Upper Deck Beams	36 The Breast Hook	
18 The Upper Tier of Ports	37 The Main Hatchway	
19 The Ladder to go on the Quarter Deck	38 The After Hatchway	
	39 The Hatchway of the Cockpit	
	40 The Main Transom	
	41 The Main Transom Knees	

Note: The draught on this page shows the same ship as in the section on hull design, but there are some surprising differences. The cutwater under the head appears to be slightly curved, a feature which became common on British ships in the mid-seventeenth century. Its appearance here may simply be due to careless drawing. The second gun-port from the bow on the upper deck is out of position. In its new position it coincides with the foremast, which would have left no room for recoil.

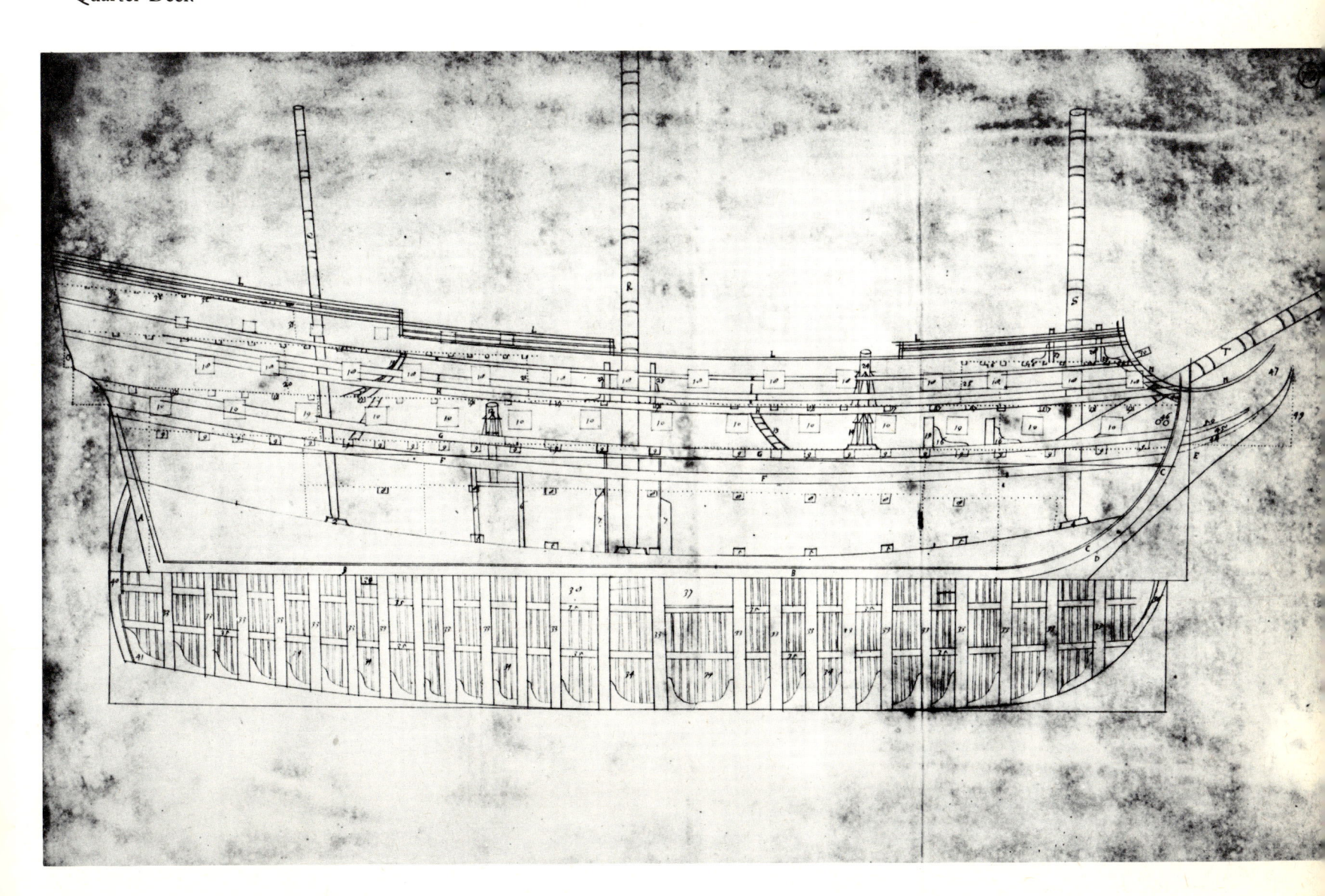

Dimensions ready calculated to draw any draught from the 6th to the 1st Rate.

p73 A good proportion for the ships of war with the true length, breadth, and depth, together with the assignment of every part of a ship to complete and finish, which with the scantling herein mentioned shall not exceed in draught of water here declared, nor want quality equal to any yet built, besides the vast charge saved, there being no alteration needed by girdling, furring, thickening, or lengthening. But being built by this proportion declared, with the exact scantling, and not erring from the way laid down in the foregoing problem, will assure the desires of the builder, and be of good service unto his Majesty, and none of the kingdom's worst subjects.

Note: In the tables on these pages the use of one figure above another refers to the depth and thickness of the piece concerned. Thus $0\text{-}\frac{12}{6}$, for the upper rail of the head of a 1st Rate, means that it will be 12 inches deep and 6 inches thick.

The following terms are not explained elsewhere:
Coamings. Ledges of timber round the edges of the hatchways, to prevent water on deck from entering.
Kevels. 'small pieces of timber nailed to the inside of the ship, unto which we belay the sheets and tacks.' (Mainwaring)
Lights. The windows of the stern cabins.
Ranges. Racks for belaying pins, which were situated 'upon the forecastle a little abaft of the foremast', and 'in the beakhead before the woldings of the bowsprit.' (Mainwaring)
Waterways. Pieces of timber round the edges of the planking next to the edge of the ship, to prevent water on deck from running onto the sides.

	1st Rate (ft ins)	2nd Rate (ft ins)	3rd Rate (ft ins)	4th Rate (ft ins)	5th Rate (ft ins)	6th Rate (ft ins)
Draught of Water	19-10	19-3	17-1	16-1	12-10	9-6
KEEL						
Length by the Keel	131-5	125-0	120-0	108-0	88-0	68-0
Depth of the Keel	1-9	1-8	1-5	1-4	1-2	1-1$\frac{1}{2}$
Breadth in the Middle of the Keel	1-10	1-9	1-6	1-3$\frac{1}{2}$	1-2	1-1
Breadth at the Post	1-2	1-1$\frac{1}{2}$	1-0$\frac{1}{2}$	0-11$\frac{1}{4}$	0-10	0-9
POST						
Length of the Post to Upper Transom	26-4	25-0	23-9	21-2	17-0	12-1
Bigness at the Head Square	2-9	2-8	2-6	2-2	0-19	1-3
Bigness below Thwartships	1-2	1-1$\frac{1}{2}$	1-0$\frac{1}{2}$	0-11$\frac{1}{4}$	0-10	0-9
Bigness below Fore and Aft	3-0	2-10	2-9	2-7	2-0	1-10
Rake of the Stem and Sweep	32-6	31-0	27-0	24-6	18-6	15-4
Bigness Thwartships	1-6	1-4$\frac{1}{2}$	1-3	1-2$\frac{1}{2}$	1-2	1-0$\frac{1}{2}$
Fore and Aft	1-10	1-8	1-5$\frac{1}{2}$	1-4	1-2$\frac{1}{2}$	1-2
WALES						
Height of the Lower Edge of the Lower Wale by Perpendicular	24-0	23-8	22-2	18-8	16-6	11-1
Height Afore to the Lower Edge Perpendicular	18-4	18-0	15-8	14-9	12-2	8-0
Height in the Midships Perpendicular	15-4	15-3	13-3	12-1	10-1	7-2
Depth of Lower Wale	1-10	1-8	1-6	1-3$\frac{1}{2}$	1-1	0-10$\frac{1}{2}$
Depth of Upper Wale	-17	-16	-14$\frac{1}{2}$	-13	-11	-10
Thickness of Upper Wale	0-10$\frac{1}{2}$	0-10	0-7$\frac{1}{2}$	0-7	0-6	0-5$\frac{3}{4}$
Depth between the Wales	2-2	2-1	2-$\frac{1}{2}$	1-9	1-3	1-2
To the lower edge Channel Wale	5-10	5-6	5-4	5-1	4-9	—
Depth of the Lower Channel Wale	1-2	1-$\frac{3}{4}$	1-0	0-10	0-8$\frac{1}{2}$	0-7
Depth between the Channel Wales	1-1	1-$\frac{1}{2}$	1-0	0-10	0-9	—
Depth of Upper Channel Wale	1-0	0-11	0-10	0-9	0-7$\frac{1}{2}$	—
Thickness of Channel Wales	0-6	0-6	0-5$\frac{1}{2}$	0-5	0-4	—
From the Channel Wales to the Great Rail	3-7	3-5	3-0	2-6	—	—
Depth of the Great Rail	0-9	0-8$\frac{1}{2}$	0-8	0-6$\frac{3}{4}$	—	—
From that to the top of the Side	3-6	3-2	2-10	2-6	—	—
Opening between the Rails	0-11	0-10	0-9	0-8$\frac{1}{2}$	0-8	—
HEAD						
Depth of the Rails	0-6	0-5$\frac{3}{4}$	0-5	0-4$\frac{3}{4}$	0-4	0-3$\frac{1}{2}$
Length of the galleries	17-0	15-6	14-3	12-4	9-6	8-0
Length of the Knee of the Head	22-6	21-6	19-8	16-6	13-6	11-0
The Steeving of the Knee	14-8	14-2	12-6	10-6	8-2	7-8
Length of the Figure	15-6	14-6	13-6	12-0	9-0	8-0
Depth of the Figure	8-0	7-6	2-10	2-4	1-10	1-6
Thickness of the Figure	3-0	2-9	2-4	1-8	1-4	—
Depth of the Trail Board	1-7	1-6	1-5	1-4	1-3	1-1
Depth of the Cheeks Upper	0-11	0-10$\frac{1}{2}$	0-9	0-9	0-8	0-7
Depth of the Lower Cheeks	1-0	0-11	0-10	0-9$\frac{1}{2}$	0-9	0-8
Upper Rail of the Head	$0\text{-}\frac{12}{6}$	$0\text{-}\frac{11}{5\frac{1}{2}}$	$0\text{-}\frac{10}{5}$	$0\text{-}\frac{9}{4\frac{1}{2}}$	$0\text{-}\frac{8}{4}$	$0\text{-}\frac{7}{3\frac{1}{2}}$
Second Rail	$0\text{-}\frac{10}{5}$	$0\text{-}\frac{9}{4\frac{1}{2}}$	$0\text{-}\frac{8}{4}$	$0\text{-}\frac{7\frac{1}{2}}{3\frac{3}{4}}$	$0\text{-}\frac{6\frac{1}{2}}{3\frac{1}{4}}$	$0\text{-}\frac{6}{3}$
Lower Rail	$0\text{-}\frac{9}{4\frac{1}{2}}$	$0\text{-}\frac{8}{4}$	$0\text{-}\frac{7\frac{1}{2}}{3\frac{3}{4}}$	$0\text{-}\frac{6}{3}$	$0\text{-}\frac{6}{3}$	$0\text{-}\frac{5}{2\frac{1}{2}}$
Kelson of the Head	1-2	1-1	1-0	0-10$\frac{1}{2}$	0-9$\frac{1}{2}$	0-8
At the Foremast End	0-10	0-9	0-8	0-7$\frac{1}{2}$	0-6$\frac{1}{2}$	0-5
Catheads Square	1-7	1-6	1-5	1-4	1-2	1-0
Stanchions of the Bulkheads Afore	0-8	0-7	0-6	0-5	0-4$\frac{1}{2}$	0-4

HULL DESIGN

	1st Rate (ft ins)	2nd Rate (ft ins)	3rd Rate (ft ins)	4th Rate (ft ins)	5th Rate (ft ins)	6th Rate (ft ins)
TIMBERS						
Floor Timbers Fore and Aft	1-3	1-2$\frac{1}{2}$	1-2	1-1	1-0	0-10
Deep upon the Keel	1-6	1-5	1-4$\frac{1}{2}$	1-4	1-1	0-10
Deep at the Rungheads	1-1	1-$\frac{1}{2}$	1-0	0-10$\frac{1}{2}$	0-9$\frac{1}{2}$	0-7
Length of the Floor Timber	26-6	25-6	23-0	19$\frac{1}{2}$-0	17-0	13-6
Room and Space	2-6	2-5	2-4	2-3	2-2	2-0
Lower Futtocks Fore and Aft	1-2$\frac{1}{2}$	1-1$\frac{3}{4}$	1-1$\frac{1}{2}$	1-$\frac{1}{2}$	1-0	0-8$\frac{1}{2}$
Upper Futtocks Fore and Aft	1-1$\frac{1}{2}$	1-1	1-0	0-11	0-10	0-7
In and Out at Breadth	0-10$\frac{1}{2}$	0-10	0-9$\frac{1}{4}$	0-8$\frac{1}{2}$	0-6$\frac{1}{4}$	0-5
Toptimbers Fore and Aft Alow	1-1	1-$\frac{1}{2}$	1-0	0-10	0-8$\frac{1}{2}$	0-7
At the Head Fore and Aft	0-9$\frac{1}{2}$	0-9	0-8$\frac{1}{2}$	0-7	0-5$\frac{1}{2}$	0-4$\frac{1}{2}$
In and Out at the Gunwale	0-4$\frac{1}{2}$	0-4	0-3$\frac{1}{2}$	0-2$\frac{1}{2}$	0-2$\frac{1}{4}$	0-2
Floor Riders Fore and Aft	1-8	1-6	1-5	1-4	1-3	1-2
At the Rungheads Up and Down	1-1	1-0$\frac{1}{2}$	1-0	0-11	0-10	0-9
Futtock Riders Fore and Aft	1-5	1-3$\frac{1}{2}$	1-3	1-1	1-0	0-11
Kelson Thwartships	1-10	1-8	1-6	1-3$\frac{1}{2}$	1-2$\frac{1}{2}$	1-1
Up and Down	1-6	1-5	1-4	1-2	1-1	1-0
BEAMS						
Orlop Beams Fore and Aft	1-5	1-4	1-3	1-2	0-0	0-0
Up and Down	1-4	1-3$\frac{1}{2}$	1-2$\frac{1}{2}$	1-1	0-0	0-0
Orlop Beams Round	0-7	0-6	0-5	0-4	0-0	0-0
Gun Deck Beams Fore and Aft	1-6	1-5	1-4	1-3	1-2	1-1
Up and Down	1-5	1-4	1-3	1-2	1-1$\frac{1}{2}$	1-0
Gun Deck Knees Up and Down	1-$\frac{1}{4}$	0-11$\frac{1}{2}$	0-10	0-8$\frac{1}{2}$	0-7$\frac{3}{4}$	0-6$\frac{1}{2}$
Carlines for the Gun Deck	0-$\frac{10\frac{1}{2}}{9}$	0-$\frac{10}{8\frac{1}{2}}$	0-$\frac{10}{8}$	0-$\frac{9}{7\frac{1}{2}}$	0-$\frac{8}{6\frac{1}{2}}$	0-$\frac{7}{6}$
Ledges of the Gun Deck	0-$\frac{6}{5}$	0-$\frac{6}{4\frac{1}{2}}$	0-$\frac{5\frac{1}{2}}{4}$	0-$\frac{5}{3}$	0-$\frac{4}{3}$	0-$\frac{3\frac{1}{2}}{2\frac{1}{2}}$
Main Bitt Pins	$\frac{2-0}{1-10}$	$\frac{1-10}{1-8}$	$\frac{1-8}{1-6}$	$\frac{1-6}{1-4}$	$\frac{1-1}{4-2}$	$\frac{1-1}{2-0}$
Fore Bitt Pins	$\frac{1-6}{1-4}$	$\frac{1-4\frac{1}{2}}{1-2\frac{1}{2}}$	$\frac{1-2}{1-1}$	$\frac{1-1}{1-0}$	$\frac{1-\frac{1}{2}}{11-0}$	0-0
Main Partners	0-11$\frac{1}{2}$	0-10$\frac{1}{2}$	0-9$\frac{1}{4}$	0-8$\frac{1}{2}$	0-7	0-6
Fore Partners	0-10	0-9	0-8	0-7	0-6$\frac{1}{2}$	0-0
Main Capstan	2-5	2-4	2-2	2-0	1-10	1-6
Gun Deck Standards Fore and Aft	1-2	1-$\frac{1}{2}$	0-11	0-10	0-8	0-7
FOOTWALING						
4 Strakes of Footwaling next the Keel	0-7$\frac{1}{2}$	0-7	0-6$\frac{1}{2}$	0-5$\frac{1}{2}$	0-5	0-4
3 Strakes at the Rungheads	0-10$\frac{1}{2}$	0-9$\frac{3}{4}$	0-9	0-8	0-6$\frac{1}{2}$	0-4
2 Strakes above them	0-9	0-8	0-7	0-6	0-4	0-3
Middlebands 2 Strakes	0-9	0-8	0-7	0-6	0-4	0-0
One Opening Strake Next	0-10	0-9	0-8	0-7	0-6	0-0
Clamps of the Gun Deck	0-10$\frac{1}{2}$	0-10	0-9	0-8	0-6	0-4
Cross Pillars in Hold	0-$\frac{13}{12}$	0-$\frac{12}{11}$	0-$\frac{10}{11}$	0-0	0-0	0-0
Knees of the Cross Pillars in Hold	0-10	0-9	0-8	0-0	0-0	0-0
Upright Pillars in Hold Square	0-9	0-8$\frac{1}{2}$	0-8	0-7	0-6	0-0
Step of the Mainmast Fore and Aft	3-2	2-10	2-6	2-4	2-2	0-0
Breast Hooks Up and Down	1-3$\frac{1}{2}$	1-3	1-2	1-1	1-0	0-9
Spirketting the Gun Deck	0-7	0-6	0-5	0-4	0-4	0-3
Plank between the Ports	0-4	0-3	0-3	0-2$\frac{1}{2}$	0-2$\frac{1}{2}$	0-0

	1st Rate (ft ins)	2nd Rate (ft ins)	3rd Rate (ft ins)	4th Rate (ft ins)	5th Rate (ft ins)	6th Rate (ft ins)
BETWEEN DECKS						
Clamps of the 2nd Deck	0-6	0-5½	0-4	0-4	0-3½	0-0
2nd Deck Beams Fore and Aft	1-2	1-2	1-½	0-10	0-8	0-0
Thickness Up and Down	1-1	1-0	0-10⅓	0-8	0-7	0-0
Knees Fore and Aft	0-9	0-8½	0-7	0-6¼	0-5½	0-0
2nd Deck Carlines	0-$\frac{7½}{9}$	0-$\frac{7}{9}$	0-$\frac{6}{8}$	0-$\frac{6}{7}$	0-$\frac{5½}{4½}$	–
2nd Deck Ledges	0-$\frac{4½}{4}$	0-$\frac{4}{3¼}$	0-$\frac{4}{3}$	0-$\frac{3¾}{3}$	0-$\frac{3}{2½}$	0-0
Standards Fore and Aft	0-10	0-9½	0-8	0-0	0-0	0-0
Plank to Lay the Deck	0-3	0-3	0-2½	0-0	0-0	0-0
Jeer Capstan	1-10	1-9	1-8	1-6	0-0	0-0
2nd DECK						
Thickness of Waterways	0-6	0-5½	0-5	0-4	0-0	0-0
Thickness of Spirketting	0-5	0-5	0-3	0-3	0-0	0-0
Clamps of the Upper Deck	0-4½	0-4½	0-4	0-3½	0-0	0-0
Plank between the ports	0-3	0-3	0-2	0-0	0-0	0-0
Upper Deck Beams Fore and Aft	1-0	0-11½	0-$\frac{6}{5}$	0-11½	0-10	0-8½
Up and Down	0-10	0-9½	0-9½	0-7⅓	0-7	0-0
Carlines of Fir	0-$\frac{7½}{5½}$	0-$\frac{7½}{5½}$	0-$\frac{8}{6}$	0-$\frac{7}{5}$	0-$\frac{6}{4½}$	0-0
Ledges of Fir	0-$\frac{4}{3½}$	0-$\frac{3¾}{3½}$	0-$\frac{3½}{3}$	0-0	0-0	0-0
Knees Fore and Aft	0-7	0-6	0-0	0-0	0-0	0-0
UPPER DECK						
Laid with Spruce Deals	0-2½	0-2½	0-0	0-0	0-0	0-0
Spirketting the Upper Deck	0-3	0-3	0-0	0-0	0-0	0-0
Waterways of the Same	0-4	0-4	0-0	0-0	0-0	0-0
Gunwales	0-4	0-4	0-3	0-3	0-2	0-2
COACH						
Beams of the Coach	0-$\frac{8}{6}$	0-$\frac{7}{5½}$	0-$\frac{6}{5}$	0-$\frac{5½}{4¾}$	0-$\frac{5}{4}$	0-$\frac{4½}{4}$
Clamps or Risings	0-$\frac{6½}{2½}$	0-$\frac{6}{2½}$	0-$\frac{5}{2½}$	0-$\frac{4}{2}$	0-$\frac{4}{1½}$	0-$\frac{4}{1½}$
Knees Fore and Aft	0-6	0-5½	0-5	0-4¾	0-4	0-4
FORECASTLE						
Forecastle Beams	0-$\frac{7¼}{5½}$	0-$\frac{7}{5½}$	0-$\frac{7}{5½}$	0-$\frac{6}{5}$	0-$\frac{4½}{4}$	0-$\frac{4}{3½}$
Clamps at Each Edge	0-$\frac{4}{2}$	0-$\frac{3½}{2}$	0-$\frac{3½}{2}$	0-$\frac{3}{2}$	0-$\frac{2}{1½}$	0-0
Knees Fore and Aft	0-6	0-5½	0-5	0-4½	0-4	0-0
Roundhouse Beams	0-$\frac{4½}{4}$	0-$\frac{4½}{4}$	0-$\frac{4}{3½}$	0-$\frac{3¾}{3}$	0-$\frac{3½}{3}$	0-0
Knees Fore and Aft	0-4½	0-4½	0-3½	0-0	0-0	0-0
BITT PINS						
Main Topsail Sheet Bitts	1-0	1-0	1-0	0-11	0-9	0-8
Jeer Bitts	1-0	1-0	0-11	0-10	0-9	0-8
Fore Topsail Sheet Bitts	0-10½	0-10	0-9	0-8	0-7½	0-7
Fore Jeer Bitts	0-10½	0-10½	0-9	0-8	0-7½	0-7
PLANKING						
Fore Bulkhead Stanchions	0-$\frac{6½}{5½}$	0-$\frac{6½}{5½}$	0-$\frac{6}{5}$	0-$\frac{5}{4½}$	0-$\frac{4½}{4}$	0-$\frac{4}{3}$
Rails of the Bulkheads	0-$\frac{4½}{3}$	0-$\frac{4½}{3}$	0-$\frac{4}{3}$	0-$\frac{3½}{2¾}$	0-$\frac{3}{2½}$	0-$\frac{3}{2}$
Plank Underwater Without Board	0-4	0-4	0-4	0-3	0-3	0-3
One Strake under the Lower Wale	0-6½	0-6	0-5	0-4	0-4	0-4
Next Strake under that	0-6	0-5	0-4	0-3	0-3	0-3
The 3rd Strake under that	0-5	0-4	0-3	0-3	0-3	0-3
Between the Wales 2 Strakes	0-6½	0-6	0-5	0-4	0-4	0-4
One Strake above the Wale	0-6	0-6	0-5	0-4	0-4	0-4

	1st Rate (ft ins)	2nd Rate (ft ins)	3rd Rate (ft ins)	4th Rate (ft ins)	5th Rate (ft ins)	6th Rate (ft ins)
2nd Strake above the Wale	0-5$\frac{1}{2}$	0-5	0-4	0-4	0-3	0-3
3rd Strake above the Wale	0-4	0-4	0-3	0-3	0-3	0-2$\frac{1}{2}$
All the rest to the Channel Wale	0-4	0-4	0-3	0-3	0-2$\frac{1}{2}$	0-2
Two Strakes above the Channel Wale	0-3	0-3	0-3	0-2$\frac{1}{2}$	0-2	0-1$\frac{1}{2}$
Two Strakes above them	0-3	0-3	0-2$\frac{1}{2}$	0-2	0-2	0-1$\frac{1}{2}$
To the Top of the Side Spruce	0-2$\frac{1}{2}$	0-2$\frac{1}{2}$	0-2	0-1$\frac{3}{4}$	0-1$\frac{1}{2}$	0-1$\frac{1}{2}$

BEAMS ROUNDING

	1st Rate	2nd Rate	3rd Rate	4th Rate	5th Rate	6th Rate
Rounding of the Orlop Beams	0-5	0-5	0-4$\frac{1}{2}$	0-4	0-8	0-7$\frac{1}{2}$
Rounding of the Gun Deck Beams	0-8	0-7$\frac{1}{2}$	0-6$\frac{1}{2}$	0-6	0-5$\frac{1}{2}$	0-0
Rounding of the 2nd Deck Beams	0-10$\frac{1}{4}$	0-10	0-0	0-0	0-0	0-0
Rounding the Upper Deck Beams	0-12$\frac{1}{2}$	0-12	0-11$\frac{1}{2}$	0-10	0-8	0-7
Rounding the Forecastle Beams	1-4	1-3	1-2$\frac{1}{2}$	1-2	1-2	1-2
Rounding the Coach Beams	1-4	1-3	1-3	1-2	1-2	1-2
Rounding the Roundhouse Beams	1-6	1-5	1-4	1-3	1-3	1-3

HATCHWAYS

	1st Rate	2nd Rate	3rd Rate	4th Rate	5th Rate	6th Rate
Main Hatchway Fore and Aft	8-6	8-4	8-2	8-0	7-6	6-6
Breadth of the Same	6-10	6-8	6-6	6-0	5-9	5-6
Breadth of the Gratings	6-6	6-4	6-0	5-9	5-2	0-0
Bigness of the Long Carlines	0-$\frac{12}{11}$	0-$\frac{12}{11}$	0-$\frac{10}{11}$	0-$\frac{9\frac{1}{2}}{10\frac{1}{4}}$	0-$\frac{9}{9\frac{1}{2}}$	0-0
Grating Ledges	0-3$\frac{1}{2}$	0-3$\frac{1}{2}$	0-3	0-3	0-3	0-0
Opening Between	0-3$\frac{1}{2}$	0-3	0-3	0-3	0-3	0-0
Quarter Deck Gratings Wide	3-10	3-8	3-6	3-0	2-6	0-0
Forecastle Gratings Wide	3-10	3-8	3-6	3-0	2-6	2-0

STERN

	1st Rate	2nd Rate	3rd Rate	4th Rate	5th Rate	6th Rate
The Main Transom	1-6	1-5	1-4	1-2	1-1	0-11
2nd Transom	1-4	1-3	1-2	1-1	0-11	0-9
3rd Transom	1-1	1-$\frac{1}{2}$	0-11	0-10	0-9	0-7
Fashion Pieces Fore and Aft	1-$\frac{1}{2}$	1-0	0-11	0-10	0-8$\frac{1}{2}$	0-7
The Main Transom Knees Up and Down	1-0	0-11	0-10	0-9	0-8	0-7
The Tiller Transom Up and Down	1-0	0-11$\frac{1}{2}$	0-9	0-8$\frac{1}{2}$	0-7$\frac{1}{2}$	0-6$\frac{1}{2}$
Knees Up and Down	0-9	0-8$\frac{1}{2}$	0-7$\frac{1}{2}$	0-7	0-6	0-5$\frac{1}{2}$
Lower Counter Rail Up and Down	1-0	0-11	0-10	0-8$\frac{1}{2}$	0-7$\frac{1}{2}$	0-6
Fore and Aft	0-9	0-8$\frac{1}{2}$	0-8	0-6$\frac{1}{4}$	0-5$\frac{1}{2}$	0-5
Rail under the Lights	0-10	0-9$\frac{1}{2}$	0-9	0-8	0-7	0-6
Fore and Aft	0-8	0-7$\frac{1}{2}$	0-7	0-6	0-5$\frac{1}{4}$	0-4$\frac{1}{2}$
Rail above the Lights Up and Down	0-9	0-8$\frac{1}{2}$	0-8	0-7	0-6	0-5
Fore and Aft	0-6	0-6	0-5$\frac{3}{4}$	0-4$\frac{3}{4}$	0-4	0-3$\frac{1}{2}$
Rails of the 2nd Light	0-$\frac{8\frac{1}{2}}{6}$	0-$\frac{8}{6}$	0-0	0-0	0-0	0-0
Fore and Aft	0-6	0-6	0-0	0-0	0-0	0-0
Depth of the Taffrail	3-2	3-0	3-8	3-6	2-4	2-0
Depth at the Ends	1-6	1-6	1-4	1-3	1-2	1-0
Thickness of the Taffrail	0-7$\frac{1}{2}$	0-7$\frac{1}{4}$	0-6$\frac{1}{4}$	0-6	0-5$\frac{1}{2}$	0-4$\frac{1}{2}$
Rounding of the Taffrail Alow	2-0	1-10	1-8	1-6	1-5	1-4
Bigness of the Terms	$\frac{2\ 0}{1\text{-}9}$	$\frac{1\ 10}{1\text{-}5}$	$\frac{1\ 8}{1\text{-}4}$	$\frac{1\ 6}{1\text{-}3}$	$\frac{1\ 5}{1\text{-}2}$	$\frac{1\ 4}{1\text{-}1\frac{1}{2}}$
Lower Counter Bracket Thwart	0-11	0-10	0-9	0-8	0-7	0-6$\frac{1}{2}$
2nd Counter Brackets Thwart	0-9	0-9	0-8	0-6$\frac{1}{2}$	0-5$\frac{1}{4}$	0-4$\frac{1}{2}$
In the Great Cabin Lights	0-7$\frac{1}{2}$	0-7$\frac{1}{2}$	0-7	0-6	0-5	0-4

	1st Rate (ft ins)	2nd Rate (ft ins)	3rd Rate (ft ins)	4th Rate (ft ins)	5th Rate (ft ins)	6th Rate (ft ins)
RUDDER						
Rudder Head Thwart	1-11	1-10	1-6	1-5	1-4	1-2½
Fore and Aft	1-10	1-10	1-6	1-5	1-5	1-3
Bigness Alow Thwartships	0-11	0-10½	0-9	0-7¼	0-6	0-5
Breadth Fore and Aft	3-10	3-9	3-8	3-6	3-5	3-5
The Tiller at the After Ends	1-4	1-3½	1-0	0-11½	0-10	0-8
At the Foremost End	0-7	0-7	0-6	0-5	0-4½	0-4
Chess Trees Fore and Aft	1-0	0-11	0-10	0-9	0-8	0-0
In and Out	1-2	1-2	1-1	1-0	0-0	0-0
HALYARD BLOCKS						
Topsail Halyard Blocks	0-11½	0-11	0-10	0-9	0-8	0-7
Ranges	0-6	0-6	0-5	0-4½	0-4	0-3½
Kevels	0-4	0-4	0-3½	0-3	0-3	0-2½
Fore Topsail Halyard Blocks	0-11	0-11	0-9½	0-9	0-8	0-7
Main Sheet Block Up and Down	1-0	1-0	0-11	0-10	0-9	0-8
Fore Sheet Block	0-11	0-11	0-9	0-8½	0-7½	0-6½
Spritsail Sheet Block	0-10	0-10	0-8½	0-8	0-7	0-6

Part 3   Rigging

PART 3. RIGGING

I do not intend to offer a full explanation of Deane's rigging plans, since there are several works on the rigging of ships of this period, especially R C Anderson's *Seventeenth Century Rigging* and Lees' *The Masting and Rigging of English Ships of War* and since Deane himself makes it clear that he is not particularly interested in the subject, it being 'a labour more proper for a master of attendant than a shipwright'. Apart from his 'preliminaries', in which he gives a method for calculating the thickness of each rope, the information given is all statistical. All the main figures are given on pages 3 and 4 of the original and the lists alongside each draught serve mainly as an explanation to the draughts. Probably they are there because Deane originally intended, as he says, to number the ropes on each draught.

It should be noted that, as well as the rigging proper, he also includes other ropes performing various functions, such as the boat ropes and the butt slings, for hoisting in casks. In each draught a curved line appears just above the keel. This is not the rising line, as described in the section on hull design, but the cutting down line, which represents the top of the kelson, and provides a resting place for the masts.

The following glossary of rigging terms used by Deane is not intended to be exhaustive, but merely to cover some of the more obscure or obsolete of them.

Boat Rope. The rope used to tow the boat astern.
Bowsing Tackle. Perhaps used for lifting heavy weights, since 'bowse' as defined as 'a word they use when they would have men pull together'. (Mainwaring's *Seamen's Dictionary*, c1620)
Bridles. Ropes used to spread the effect of the bowline over the edge of the rail.
Breast Rope. 'The ropes which make fast the parrel to the yard.' (Mainwaring)
Burtons. Tackle used to lift heavy weights.
Catharpins. 'The use whereof is to force the shrouds tauter for the better ease and safety of the mast in the rolling of the ship.' (Mainwaring)
Cat Rope. The rope used to hoist the anchor up to the cat-head.
Collar. The rope to which the main stay is fastened.
Fish Tackle. Used to hold up the fluke of the anchor, when the other end was held by the cat rope.
Gust Rope, or *Guest Rope.* Used in combination with the boat rope, 'to keep the boat from sheering' (Mainwaring). Deane uses both spellings, whereas Mainwaring calls it the 'gestrope'.
Knave Line. Used to 'keep the ties and halyards from turning about one another'. (Mainwaring)
Luff Tackle. Probably a rope used as an aid to the tack in strong winds.
Passing Rope. Probably the same as *Passarado*, 'any rope wherewith we haul down the sheet blocks of the main and fore sails when they are hauled aft.' (Mainwaring)
Robins. Small ropes used to fasten the head of the sail to the yard.
Shank Painter. Used to hold up the stock of the anchor while the ends were held by the cat rope and the fish tackle.
Stoppers. Mainly used to hold the anchor cable in position when coming to anchor.
Viol. Used for lifting the anchor when it is stuck fast, by attaching one strand of the cable to the jeer capstan to give extra purchase.
Wolding. Ropes used to hold the bowsprit to the bows.
Yard Lines. Possibly ropes under the yards to support the men working on the sails.

DEANE'S DOCTRINE OF NAVAL ARCHITECTURE, 1670

The masts and yards ready calculated, and how to find the lengths of the same.

	1st Rate (yds ins)	2nd Rate (yds ins)	3rd Rate (yds ins)	4th Rate (yds ins)	5th Rate (yds ins)	6th Rate (yds ins)
Length of the Mainmast	36-12	33-12	31-12	29-0	26-12	21-4
Bigness at the Partners	0-36$\frac{1}{2}$	0-33$\frac{1}{4}$	0-29$\frac{1}{2}$	0-27	0-21	0-17
Bigness at the Top	0-24$\frac{1}{3}$	0-22$\frac{1}{2}$	0-19	0-18	0-14	0-11$\frac{1}{2}$
Foremast Long	29-0	28-0	27-12	26-4	22-12	18-24
Bigness of the Partners	0-29	0-28	0-27$\frac{1}{2}$	0-24$\frac{1}{2}$	0-18	0-17
Bigness at the Top	0-19	0-18$\frac{2}{7}$	0-18	0-16	0-12	0-11$\frac{7}{8}$
Length of the Bowsprit	22-8	21-8	21-6	0-18	16-26	14-0
Bigness in the Partners	0-32	0-31	0-29	0-26	0-20	0-16
Bigness at the Top	0-16	0-15$\frac{3}{4}$	0-14$\frac{1}{2}$	0-13	0-10	0-8
Length of the Mizzen Mast	26-12	25-24	25-0	23-5	20-24	16-27
Bigness in the Partners	0-21	0-20	0-18	0-16	0-13	0-10$\frac{1}{2}$
Length of the Main Topmast and bigness as above proposed, and to be observed as the rest	21-19	20-24	19-18	17-0	15-12	12-0
Length of the Main Topgallant Mast	9-12	8-30	8-0	7-0	6-0	4-18
Length of the Fore Topmast	18-18	18-0	17-0	15-0	13-0	10-18
Length of the Fore Topgallant Mast	8-18	8-8	7-21	6-0	5-0	4-0
Length of the Spritsail Topmast	7-18	7-4	6-0	5-0	4-0	0-0
Length of the Mizzen Topmast	11-0	10-18	9-18	9-0	8-0	5-18
Length of the Main Yard	33-18	32-6	28-0	26-10	22-9	15-2
Length of the Fore Yard	29-24	27-12	25-10	22-4	19-24	13-4
Length of the Spritsail Yard	19-21	18-12	16-18	15-0	13-0	10-0
Length of the Mizzen Yard	26-24	26-18	25-12	22-4	19-24	13-4
Crossjack Yard	17-21	17-12	16-18	15-0	13-0	10-0
Mizzen Topsail Yard	8-29	8-24	8-9	7-18	6-18	5-0
Main Topsail Yard	18-22	17-16	16-4	14-12	11-10	9-14
Fore Topsail Yard	16-12	14-28	13-0	12-4	10-10	7-29
Spritsail Topsail Yard	8-29	8-24	8-9	7-18	6-18	5-0
Fore Topgallant Yard	7-18	7-2	6-18	6-2	5-5	4-21
Main Topgallant Yard	8-29	8-24	8-4	7-18	6-0	4-21
Stormsail Booms	16-18	16-0	15-0	13-18	11-4	9-22

p75 The most exact way to find out the length of any ship's mainmast, which being once found is the rule from whence other masts is derived, will be shown. As suppose the ship's keel to be 120 feet in length, her breadth 36 feet, and the depth half breadth, 18 feet; which add all together, makes 174. This product 174 divide by 5 as thus — 34$\frac{4}{5}$. This quotient is 34$\frac{4}{5}$ yards, but by reason the ship's breadth is more than 27 feet, you must deduct it out of the quotient, 34$\frac{4}{5}$ yards, and the remainder is the true length required, as by deducting 9 feet or 3 yards, the ship being so much more than 27 feet in breadth, out of the 34$\frac{4}{5}$ yards, and there will remain 31$\frac{4}{5}$ yards, which is an exact proportion and suitable length for such a ship. Only, by the way, observe that if the ship be under 27 feet broad, then add so many feet to the quotient as will make it 27 feet. As for example, a ship 66 feet by the keel and 21 feet broad, and the half breadth 10$\frac{1}{2}$ feet. Add them all together, 29$\frac{1}{2}$, which product is thus 19$\frac{2}{5}$ yds. But by reason the ship is two yards less in breadth than 27 feet, you must add 6 feet or two yards to the quotient to make it 27 feet, the ship being less than 27, and the quotient will then be 21$\frac{2}{5}$ yards, which is a fit length for the mainmast of a ship of these dimensions. Having now showed you two examples which hold good in all ships of war built by my method, I shall leave you to this, knowing many other ways of doing this matter, but none so good, by reason this respects length, breadth and depth, and is most exact of all others which I ever yet could find. Having now found out the length of the mainmast, I allow for every yard the mast is long $\frac{15}{16}$ of an inch for its diameter in the partners[1], and then it will be 29$\frac{3}{5}$ inches in breadth. But if a New England tree, then I allow to every yard in length one inch diameter in the partners for great ships, but for lesser ships $\frac{7}{8}$ an inch or $\frac{2}{3}$ of its bigness in the head. And for my yards, I allow for every yard in length $\frac{5}{8}$ of an inch for the bigness in the slings, and $\frac{1}{3}$ of that for the bigness in the ends. Having thus made you sensible how big the masts and yards must be to their length, and the breadth of the mainmast found, shall proceed to finding out the lengths of all the rest of the masts and yards, which are as follows. The mainmast 31$\frac{2}{3}$ yds, the foremast $\frac{9}{10}$, of the mainmast. The bowsprit $\frac{2}{3}$ of the mainmast, the mizzen mast $\frac{25}{27}$ of the mainmast, the main topmast $\frac{19}{31}$ of the mainmast, fore topmast $\frac{17}{19}$ of the main topmast, main topgallant mast $\frac{8}{19}$ of the main topmast, the fore topgallant mast $\frac{7}{17}$ of the fore topmast. Mizzen topmast the $\frac{1}{2}$ of the main topmast, the spritsail topmast the $\frac{1}{3}$ of the main topmast, the main yard the length of the foremast, the fore yard $\frac{25}{28}$ of the main yard, the main topsail yard $\frac{16}{19}$ of the main topmast, the fore topsail $\frac{13}{16}$ of the main topsail, the spritsail yard the length of the fore topmast, the mizzen yard the length of the fore yard, the crossjack yard the length of the spritsail yard, and the spritsail topsail yard $\frac{1}{2}$ of the spritsail yard. Fore topgallant yard $\frac{1}{2}$ of the fore topsail yard, the main topgallant yard $\frac{1}{2}$ of the main topsail yard. These dimensions truly observed will hold good in every part. And thus I shall leave you to use as you see cause.

1. *Partners.* In the original this is spelt 'prtnrs'. This may suggest that it was written down by one of Deane's clerks, for it is an abbreviation which is far more likely to occur to a clerk than a shipwright. Possibly it was dictated, which could help account for the conversational style.

Preliminaries: Before we proceed unto the rigging of the ships, it is necessary we lay down a foundation for the sizes, which we will derive from the several masts and yards, wherein their property for strength is most suitable. And beginning with the mainmast, as for example a 1st Rate ship whose mast is 34 inches diameter. The bigness of the stay is required.

The bigness of the main stay's circumference shall be the half of the diameter of the mainmast, so it will prove 17 inches circumference, the mainmast being 34 inches diameter. The half is 17, and so for all ships of war the collar as 12 to 17.

The main shroud's circumference, the one half of the main stay's circumference, which will prove to be $8\frac{1}{2}$ inches. The pendants of tackles, the gun slings, the great buoy ropes, also of the same size of the main shrouds.

The fore stay the half of the foremast's diameter. Then will the fore stay's circumference be $15\frac{1}{2}$ inches, the foremast being 31 inches diameter, the collar as 10 to 14.

The fore shroud's circumference the half of the fore stay's circumference, which will prove $7\frac{3}{4}$ inches. The pendants of tackles, main topsail sheets, and pendants of fish tackles of the same size.

The pendant of garnet and the main sheets as 28 to 31, or 7 inches circumference to $7\frac{3}{4}$ inches, the fore shrouds.

The fore tacks, pendant of top rope, main jeers, main topsail tie, shank painters, seizing for great blocks, all in proportion unto the fore shrouds, as 30 to 31 or $7\frac{1}{2}$, the circumference to the shrouds $7\frac{3}{4}$.

The fore runners of tackles, fore jeers, fore topsail tie, fore topsail sheets, main runners of tackles, main breast rope, mizzen stay, all in proportion to the main shrouds, as 25 to 34, and is more than $\frac{2}{3}$ of the main shroud's circumference.

The fore topmast stay, main topmast runner, cat ropes, and fore sheets as 23 to 34, which is little more than $\frac{2}{3}$ of the main shrouds.

The pinnace rope, spritsail slings, horse of the head, fore topmast stay, lanyard of main stay, main parrel ropes, main guy, standing backstays, top rope fall, stoppers, stream anchors, gust rope, butt slings, all in proportion unto the main shrouds, as 21 to 34, or near $\frac{2}{3}$ of the main shrouds.

The pendants of spritsail sheets, fall of the fore top rope, fore topmast runner of halyards, fore topmast standing backstays, main topmast shrouds, mizzen jeers, shrouds, all in proportion to the main shrouds as 19 to 34, which is something than half of the same.

The lanyard of fore stay, fore parrel rope, fore bowlines, fore topmast shrouds, fall of garnet, main topsail futtocks, fall of the fish hook rope, in proportion to the main shrouds as 17 to 34, the one half of the same, as $4\frac{1}{4}$ to $8\frac{1}{2}$.

The spritsail sheets, pendants of braces, main tackle falls, lanyards of main shrouds, bowlines and bridles, pendants of main topmast tackle, mizzen sheets, mizzen parrel ropes, main lifts, braces, clew garnets, crossjack slings in proportion unto the fore shrouds as 15 to 31, which is very near one half of the fore shrouds.

The spritsail lifts, halyards, fore tackle fall, fore lifts, fore braces, fore topmast fall of halyards, all in proportion unto the fore shrouds as 14 to 31 or as $3\frac{1}{2}$ to $7\frac{3}{4}$ circumference.

The fore clew garnets, bowlines, braces, clewlines, main buntlines, main topsail lifts, mizzen truss, brails, lanyards, mizzen topmast shrouds, in proportion to the main shrouds as 14 to 34, or $3\frac{1}{4}$ to 8_2 circumference, which is near one third.

The spritsail buntlines, foretop main buntlines, fore topsail lifts, crossjack braces, all in proportion unto the main shrouds as 9 to 34, or as $2\frac{1}{3}$ to $8\frac{1}{2}$, which is little more than $\frac{1}{4}$ of the circumference of the main shrouds.

The spritsail falls of backstays, lifts, fore topgallant shrouds, braces, bowlines, bridles, clewlines, lifts, in proportion to the main shrouds, as 7 to 34, or $1\frac{3}{4}$ to $8\frac{1}{2}$, which is near $\frac{1}{5}$ circumference of the main shrouds.

The spritsail topmast shrouds, fore topgallant lanyards of shrouds, bowlines, mizzen topsail bridles, in proportion as 6 to 34 of the main shrouds, or as $1\frac{1}{2}$ to $8\frac{1}{2}$, which is a little above $\frac{1}{6}$ of the circumference of the main shrouds.

p1 In the following works you will find a most exact way for the rigging and completing of any of His Majesty's ships of war, which are of the dimensions laid down in six several demonstrations, wherein is showed the true length of every rope, and proper place of every mast and yard, together with the lengths most suitable unto their burthens and strength, from which demonstrations are set down the undeniable length of every rope. Unto which I have added and allowed for every shroud passing about the head of the mast and deadmen eyes, and length to sieze where it is proper. Also allowance is made for every rope for passing about everything in the whole ship, or for heaving at the capstan or the like. And in short all allowance whatever can be required, or pretended unto by any, as may be discovered, against every draught from the 6th to the 1st Rate, where every rope is made, and in the 1st Rate marked to know his place where I refer to for better satisfaction, concluding no denial will be made against a demonstration, as more plainly appears in the following labours. Yet, by the way, I must not omit one observation, that although I have allowed just every true length in his proper place, I affirm that the owner may receive a benefit of more than one ninth part of all the whole rigging, by the stretching of every rope ere it be put up, especially such cordage as is made in His Majesty's yards, and, if so, I conceive it here a matter worthy of remembrance. But it being none of my business, hasten to my other work of a shipwright, which is the ground design.

In the next side, where you find the several columns, are collected and set forth by way of abstract in the whole rigging and its weight, fit for the several ships in this book, and also the most exact way of size which is fit for every rope's place. And, if anyone its perusal make objection against the size, it being a matter so variously performed in rigging of ships, my answer is that every rope holds its proportion as you have been taught, from the greatest unto the smallest. But perhaps some will object and say ropes are not made of so many several sizes, which I answer is no fault in my collectings[1], there being many very near them. Nor would I have taken the pains, which indeed is a labour more proper for a master of attendant[2] than a shipwright, had not I aimed at one perfection for until such time as we know all weights overhead whatsoever, we are in the dark, which is the ground of these pains, to receive light. So that if any shall hap to view those who pretend to things of this nature, I desire a favourable censure until a demonstration appears of theirs to disprove these. The ground of my request, for as yet I never see anything either of this method or perfection which is not to declare unto the world but my own satisfaction.

The several sizes, numbers of fathoms, and total weight to rig any of His Majesty's ships, from the 6th to the 1st Rate.

p2 Note that upon this stands the several sizes, the number of fathoms and weight of every size which will rig and complete the six several ships in this book demonstrated, together with the total weight each ship requires, ready cast up in tons, hundreds, quarters and pounds, so that by your inspection you will readily find the full quantity each ship requires of the stocks, not only for the standing and running rigging but also the cat ropes, stoppers, painters, boat ropes, buoy ropes, strapping of blocks, woldings, siezings and the like, as more fully appears when every rope is demonstrated against each draught.

---

1. *Within my collectings.* That is, under his control.
2. *Master of attendant.* The officer in a Royal Dockyard responsible for moorings, ships laid up in ordinary, and the rigging of ships.

RIGGING

6th RATE

Size (ins)	Fathoms Number	Weight Tons	Cwt	Qr	Lbs
$\frac{7}{8}$	441		1	2	13
1	742		3	0	24
$1\frac{1}{8}$	209		1	0	14
$1\frac{3}{4}$	424		2	2	12
2	330		6	2	1
$2\frac{1}{8}$	204		3	2	8
$2\frac{1}{3}$	256		4	3	20
$2\frac{1}{2}$	400		7	2	23
$2\frac{5}{8}$	87		1	2	4
$2\frac{3}{4}$	100		2	0	1
3	85		1	1	18
$3\frac{3}{4}$	157		4	1	14
4	92		4	0	20
$4\frac{1}{4}$	137		6	1	27
$7\frac{1}{2}$	11		1	1	4
$6\frac{1}{2}$	4		0	1	5
6	5		0	1	12
$6\frac{1}{2}$	8		0	2	6
0	0		0	0	0
0	0		0	0	0
Total		2	13	3	2

4th RATE

Size (ins)	Fathoms Number	Weight Tons	Cwt	Qr	Lbs
$1\frac{1}{8}$	670		3	2	7
$1\frac{1}{4}$	1088		5	3	19
2	417		3	3	22
$2\frac{1}{8}$	714		9	3	7
$2\frac{3}{4}$	789		13	0	14
$3\frac{1}{8}$	309		7	1	10
$3\frac{1}{4}$	200		5	0	1
$3\frac{1}{2}$	96		2	2	5
$3\frac{1}{2}$	188		5	0	9
$3\frac{3}{4}$	255		5	1	24
$4\frac{1}{2}$	193		9	1	10
$5\frac{1}{8}$	176		13	0	10
$5\frac{1}{2}$	137		6	3	12
$5\frac{3}{4}$	253		18	2	20
6	168		12	3	20
$6\frac{1}{4}$	216		20	1	6
8	3		0	1	14
10	7		1	1	16
11	11		2	2	6
$13\frac{1}{2}$	14		4	3	25
Total		7	1	2	4

2nd RATE

Size (ins)	Fathoms Number	Weight Tons	Cwt	Qr	Lbs
$1\frac{3}{8}$	885		4	2	8
$1\frac{1}{2}$	1664		9	3	4
$2\frac{1}{4}$	694		9	0	17
$2\frac{1}{2}$	939		15	2	17
$3\frac{1}{8}$	1158		26	1	8
$3\frac{1}{2}$	422		10	2	17
4	268		12	0	21
$4\frac{1}{4}$	142		6	3	17
$4\frac{1}{2}$	302		15	1	17
5	353		20	3	27
$5\frac{1}{2}$	262		16	2	7
6	221		19	3	7
$6\frac{1}{2}$	177		15	3	14
7	365		40	1	17
$7\frac{1}{2}$	234		30	1	16
8	275		43	2	20
9	4		0	2	12
$12\frac{1}{2}$	9		2	3	4
14	14		5	1	0
16	18		9	0	0
Total		15	15	3	26

5th RATE

Size (ins)	Fathoms Number	Weight Tons	Cwt	Qr	Lbs
1	559		2	3	9
$1\frac{1}{8}$	935		5	0	4
$1\frac{1}{2}$	355		2	3	17
$1\frac{9}{10}$	581		4	3	17
$2\frac{1}{2}$	531		8	3	11
$2\frac{7}{8}$	279		6	1	9
3	168		3	3	24
$3\frac{1}{8}$	69		1	1	26
$3\frac{1}{4}$	166		4	0	2
$3\frac{1}{3}$	220		5	1	14
4	165		6	2	14
$4\frac{3}{8}$	137		5	2	17
$4\frac{1}{2}$	108		4	2	10
$4\frac{3}{4}$	225		11	3	3
5	133		8	0	11
$5\frac{3}{8}$	177		12	1	16
7	3		0	1	9
8	10		1	1	0
8	5		0	2	14
$11\frac{1}{2}$	13		3	1	16
Total		5	0	1	9

3rd RATE

Size (ins)	Fathoms Number	Weight Tons	Cwt	Qr	Lbs
$1\frac{1}{4}$	773		3	3	15
$1\frac{3}{8}$	1330		8	3	17
$2\frac{1}{8}$	487		5	3	24
$2\frac{3}{8}$	844		13	3	17
$3\frac{1}{10}$	928		19	2	26
$3\frac{2}{5}$	358		9	2	27
$3\frac{3}{4}$	143		4	3	14
4	122		5	1	4
$4\frac{1}{8}$	224		9	1	12
$4\frac{1}{4}$	305		12	2	10
$4\frac{7}{8}$	225		11	2	9
$5\frac{1}{2}$	203		14	0	12
6	164		10	1	20
$6\frac{1}{3}$	309		27	0	12
$6\frac{5}{8}$	198		18	1	7
7	254		28	1	11
9	3		0	2	0
11	8		1	3	16
12	13		3	1	10
14	18		6	3	0
Total		10	16	2	11

1st RATE

Size (ins)	Fathoms Number	Weight Tons	Cwt	Qr	Lbs
$1\frac{1}{2}$	1198		6	0	19
$1\frac{3}{4}$	1890		11	0	5
$2\frac{1}{3}$	749		9	3	12
$2\frac{3}{4}$	1189		22	0	10
$3\frac{1}{4}$	1300		30	1	7
$3\frac{3}{4}$	445		12	2	12
$4\frac{1}{4}$	333		16	0	27
$4\frac{1}{2}$	176		8	3	7
$4\frac{3}{4}$	345		16	2	17
$5\frac{1}{4}$	432		28	2	3
$5\frac{3}{4}$	225		15	3	14
$6\frac{1}{3}$	238		22	1	7
7	187		20	1	4
$7\frac{1}{2}$	255		50	2	12
$7\frac{3}{4}$	323		46	1	20
$8\frac{3}{4}$	412		71	1	27
10	5		1	0	1
13	10		3	0	24
$14\frac{1}{2}$	18		7	1	16
17	20		11	1	21
Total		20	12	1	13

The names of every rope allowed against the sizes, number of fathoms, and particular weights, to rig from the 6th to the 1st Rate.

p3 Here are collected what sizes, number of fathoms and their weights will serve to rig the six demonstrations following, from the 6th to the 1st Rate.

	6th Rate	5th Rate	4th Rate	3rd Rate	2nd Rate	1st Rate
Size (ins)	$4\frac{1}{2}$	$5\frac{3}{8}$	$6\frac{1}{4}$	7	8	$8\frac{1}{2}$
Main Shrouds (fathoms)	76	114	152	182	205	298
Buoy Ropes	50	50	50	55	60	90
Gun Slings	4	4	$4\frac{1}{2}$	5	5	7
Pendant of the Main Tackle	3	4	4	5	7	8
Pendants of Shroud Tackles	4	5	6	7	8	9
Total Weight (cwt qrs lbs)	6-1-27	12-1-16	20-1-16	28-1-11	46-1-20	71-1-27
Size (ins)	4	5	6	$6\frac{3}{4}$	$7\frac{1}{2}$	$7\frac{3}{4}$
Pendant of the Fore Tackle (fathoms)	4	5	$5\frac{1}{2}$	$6\frac{1}{2}$	$7\frac{1}{2}$	8
Fore Shrouds	56	80	112	136	168	210
Main Topsail Sheets	28	42	44	48	50	92
Pendant of the Fish Tackle	4	6	7	8	9	13
Total Weight (cwt qrs lbs)	4-0-20	8-0-11	12-3-20	18-1-7	30-1-16	40-1-20
Size (ins)	$3\frac{3}{4}$	$4\frac{3}{4}$	$5\frac{3}{4}$	$6\frac{1}{3}$	7	$7\frac{1}{2}$
Strapping of Blocks and Fore Pendants	26	34	46	56	90	98
Fore Tacks	18	27	28	32	34	42
Pendant of the Fore Top Rope	9	11	12	13	14	$14\frac{1}{2}$
Main Jeers	39	60	66	96	100	115
Main Top Tie	6	6	$7\frac{1}{2}$	9	9	10
Shank Painters	12	15	15	15	18	24
Seizing for Blocks	20	30	35	40	50	60
Main Topsail Sheets	28	42	44	48	50	92
Total Weight (cwt qrs lbs)	4-1-14	11-3-3	18-2-20	27-0-12	40-1-17	50-2-12
Size (ins)	$3\frac{1}{2}$	$4\frac{1}{2}$	$5\frac{1}{2}$	6	$6\frac{1}{2}$	7
Wolding for the Bowsprit	40	50	68	88	90	96
Pendant of the Garnet	5	8	9	10	11	11
Main Sheets	40	50	60	66	76	80
Total Weight (cwt qrs lbs)	1-1-18	4-2-10	6-3-12	10-1-20	15-3-14	20-1-4
Size (ins)	$2\frac{5}{8}$	$4\frac{3}{8}$	$5\frac{1}{8}$	$5\frac{1}{2}$	6	$6\frac{1}{3}$
Fore Runners of Tackles	16	21	22	25	28	30
Fore Jeers	20	40	70	82	90	96
Fore Topmast Tie	5	$6\frac{1}{2}$	7	7	$7\frac{1}{2}$	$9\frac{1}{2}$
Fore Topmast Sheets	21	34	36	44	45	46
Main Runners of Tackles	16	24	26	28	32	36
Main Breast Rope	2	3	4	4	5	6
Mizzen Stay	$7\frac{1}{2}$	9	11	13	14	15
Total Weight (cwt qrs lbs)	1-2-4	5-2-17	13-0-10	14-0-12	19-3-7	22-1-7
Size (ins)	$2\frac{1}{2}$	4	$4\frac{1}{2}$	$4\frac{7}{8}$	$5\frac{1}{2}$	$5\frac{3}{4}$
Horse for the Main Yard and Strapping of Blocks	33	39	50	71	85	60
Fore Topmast Stay	10	11	15	$15\frac{1}{2}$	16	19
Main Topmast Runner	10	11	12	13	18	22
Cat Ropes	30	40	43	45	50	50
Fore Sheets	36	48	54	60	70	74
Runners of the Shroud Tackles	12	16	19	21	22	24
Total Weight (cwt qrs lbs)	1-2-22	6-2-14	9-1-10	11-2-9	16-2-7	15-3-14

RIGGING

	6th Rate	5th Rate	4th Rate	3rd Rate	2nd Rate	1st Rate
Size (ins)	$2\frac{1}{3}$	$3\frac{1}{3}$	$3\frac{3}{4}$	$4\frac{1}{4}$	5	$5\frac{1}{4}$
Pinnace Rope	20	26	28	30	40	46
Slings for the Bowsprit	2	3	$3\frac{1}{2}$	4	5	8
Horse for the Head	$4\frac{1}{2}$	6	$6\frac{1}{2}$	$7\frac{1}{2}$	8	9
Lanyard of the Main Stay	6	10	12	13	14	15
Main Parrel Rope	8	10	12	14	19	25
Main Guy	7	8	8	9	9	10
Main Topmast Standing Backstays	26	64	75	108	120	156
Main Top Rope Fall	20	36	38	42	48	61
Stoppers of the Stream Anchor	6	7	9	10	10	11
Buoy Rope of the Stream Anchor	14	15	15	15	15	16
Gust Rope	0	28	28	30	40	46
Total Weight (cwt qrs lbs)	1-2-6	5-1-14	5-1-24	12-2-10	20-3-27	28-2-3
Size (ins)	5	5	5	5	5	5
Butt Slings	5	6	6	7	9	10
Pendants for the Spritsail Sheets	0	0	0	0	8	8
Fall of the Fore Topmast Top Rope	30	32	34	36	44	48
Fore Topmast of the Halyards	8	9	10	14	16	19
Fore Topmast Standing Backstays	21	55	56	66	114	126
Main Topmast Shrouds	34	48	58	76	84	104
Mizzen Jeers	20	22	30	32	36	40
Total Weight (cwt qrs lbs)	3-1-14	4-0-2	5-1-9	9-1-12	15-1-17	16-2-17
Size (ins)	$2\frac{1}{2}$	$3\frac{1}{8}$	$3\frac{1}{2}$	4	$4\frac{1}{4}$	$4\frac{1}{2}$
Main Bowlines	22	28	34	38	48	58
Mizzen Shrouds	36	41	62	84	94	128
Total Weight (cwt qrs lbs)	0-2-12	1-1-26	2-2-5	5-0-4	6-3-17	8-3-7
Size (ins)	$2\frac{1}{3}$	3	$3\frac{1}{4}$	$3\frac{3}{4}$	4	$4\frac{1}{4}$
Lanyard of the Fore Stay	4	7	8	10	12	14
Fore Parrel Ropes	7	7	$7\frac{1}{2}$	12	$14\frac{1}{2}$	16
Fore Bowlines	23	27	32	46	56	64
Fore Topmast Shrouds	30	42	58	61	62	90
Main Fall of the Garnet	24	25	27	34	34	43
Main Topmast Futtocks	16	20	22	28	34	50
Fall of the Fish Hook Rope	18	26	28	30	32	32
Buoy Rope for the Kedge Anchor	8	10	14	14	14	16
Total Weight (cwt qrs lbs)	1-2-6	3-3-24	5-0-1	4-3-26	12-0-21	16-0-27

	6th Rate	5th Rate	4th Rate	3rd Rate	2nd Rate	1st Rate
Size (ins)	4	4	4	4	4	4
Hogshead Slings	4	4	4	8	8	8
Size (ins)	$2\frac{1}{8}$	$2\frac{7}{8}$	$3\frac{1}{8}$	$3\frac{3}{5}$	$3\frac{1}{2}$	$3\frac{3}{4}$
Spritsail Fall for the Sheets	18	50	54	58	60	68
Fore Pendant of Braces	4	4	$4\frac{1}{2}$	5	6	7
Main Falls of Tackles	50	56	60	62	115	124
Main Lanyards for Shrouds	30	40	46	60	76	98
Main Bridles	8	10	10	12	16	22
Main Topmast Pendants of Tackles	0	5	6	6	6	6
Main Topmast Bowlines	36	40	48	50	50	58
Mizzen Sheets	14	15	16	18	20	21
Mizzen Parrel Ropes	4	6	6	6	7	7
Foremast Lanyard to the Shrouds	26	30	36	56	60	64
Crossjack Slings	3	4	5	5	5	6
Lanyard for the Stoppers at the Bitts	6	7	8	10	11	12
Luff Tackle and Pendant of Mizzen Burton	9	10	10	17	18	20
Total Weight (cwt qrs lbs)	3-2-8	6-1-9	7-1-10	9-2-27	10-2-17	12-2-12
Size (ins)	2	$2\frac{1}{2}$	$2\frac{3}{4}$	$3\frac{1}{10}$	$3\frac{1}{8}$	$3\frac{1}{4}$
Spritsail Lifts	12	16	18	42	50	58
Spritsail Pendants of Braces	3	4	4	4	4	5
Spritsail Halyards	12	20	22	22	27	38
Fore Tackle Falls	44	46	52	60	150	161
Fore Lifts	34	46	48	54	58	66
Fore Braces	26	34	46	50	52	58
Fore Topmast Lanyard to the Stay	6	10	10	10	10	12
Fore Topmast Falls of Halyards	18	34	35	36	42	62
Fore Topmast Pendants of Braces	3	4	4	4	4	5
Main Lifts	42	44	54	66	70	74
Main Falls of Braces	40	44	60	64	71	81
Main Clew Garnets	41	44	46	58	66	86
Main Topmast Lanyards of the Stay	4	8	8	9	11	12
Main Topmast Pendants of Braces	3	4	5	5	6	6
Total Weight (cwt qrs lbs)	6-1-1	8-1-11	9-2-14	15-2-26	20-1-8	23-1-7
Size (ins)	2	$2\frac{1}{2}$	$2\frac{3}{4}$	$3\frac{1}{10}$	$3\frac{1}{8}$	$3\frac{1}{4}$
Woldings for the Mainmast and Standing Lifts	4	6	183	211	256	262
Main Topsail Bridles	6	9	10	14	24	26
Main Topsail Clewlines	44	60	74	76	87	92
Main Topmast Fall of Halyards	36	46	50	58	58	60
Main Topmast Parrel Ropes	5	5	6	8	10	12
Mizzen Bowlines	7	9	9	10	11	16
Crossjack Lifts	5	6	7	8	10	12
Tackle Falls for Main Shrouds	26	32	48	59	80	96
Total Weight (cwt qrs lbs)	0-1-0	0-3-0	3-2-0	4-0-0	6-0-0	7-0-0
Size (ins)	$1\frac{1}{2}$	$1\frac{9}{10}$	$2\frac{1}{8}$	$2\frac{3}{8}$	$2\frac{1}{2}$	$2\frac{3}{4}$
Spritsail Clewlines	18	24	30	34	39	48
Spritsail Garnets	0	24	40	48	55	58
Spritsail Falls for Braces	28	30	40	46	47	58
Spritsail Topmast Shrouds	8	12	14	16	18	26
Spritsail Topmast Tie	2	2	2	3	3	4
Spritsail Topmast Futtocks	0	0	0	8	10	12
Fore Clew Garnets	34	44	46	52	57	68
Fore Topmast Futtocks	6	20	24	30	48	50

RIGGING

	6th Rate	5th Rate	4th Rate	3rd Rate	2nd Rate	1st Rate
Fore Topmast Bowlines	40	48	50	58	62	80
Fore Topmast Falls of Braces	30	40	44	54	54	58
Fore Topmast Clewlines	40	56	64	66	74	78
Fore Topmast Parrel Ropes	3	5	6	7	8	10
Fore Topgallant Mast Tie	2	2	$2\frac{1}{2}$	3	$3\frac{1}{2}$	4
Main Buntlines	38	50	58	78	80	86
Main Topmast Lifts	40	54	60	62	66	69
Main Topmast Fall of Braces	42	48	60	64	68	78
Main Topsail Buntlines	18	30	32	44	46	54
Main Topgallant Tie	2	3	3	3	4	6
Mizzen Lanyards for the Stay	2	2	$2\frac{1}{2}$	3	4	6
Mizzen Truss	10	12	16	26	28	36
Mizzen Brails	40	44	68	80	100	180
Mizzen Topmast Shrouds	12	14	20	26	30	36
Mizzen Topmast Futtocks	3	9	9	9	9	18
Mizzen Lanyards for Shrouds	6	8	14	24	26	32
Total Weight (cwt qrs lbs)	2-2-12	4-3-17	9-3-17	13-3-17	15-2-17	22-0-10
Size (ins)	$1\frac{1}{8}$	$1\frac{1}{2}$	2	$2\frac{1}{8}$	$2\frac{1}{4}$	$2\frac{1}{3}$
Spritsail Buntlines	18	20	25	28	37	61
Spritsail Topmast Pendants for Braces	2	3	3	3	3	3
Spritsail Topmast Pendants of Backstays	0	3	3	4	$5\frac{1}{2}$	6
Spritsail Topmast Parrel Rope	1	1	1	2	$3\frac{1}{2}$	4
Fore Buntlines	34	46	60	62	68	69
Fore Topmast Lanyards of Shrouds	7	8	12	19	20	28
Fore Topmast Lanyard of the Backstay	3	6	8	9	14	18
Fore Topsail Bridles	0	10	10	10	11	13
Fore Topsail Buntlines	0	28	30	32	34	35
Fore Topgallant Futtocks	0	0	0	4	6	8
Fore Topgallant Stay	11	14	$15\frac{1}{2}$	16	$19\frac{1}{2}$	28
Fore Topgallant Parrel Ropes	0	1	$1\frac{1}{2}$	$1\frac{1}{2}$	2	4
Main Topmast Lanyards of the Shrouds	14	15	16	26	28	40
Main Topgallant Shrouds	10	12	14	15	16	20
Main Topgallant Futtock	4	12	12	12	12	12
Main Topgallant Parrel Rope	$0\frac{3}{4}$	1	2	3	3	4
Main Topgallant Stay	8	16	17	$19\frac{1}{2}$	20	21
Fore Topsail Lifts	28	48	50	54	58	68
Main Topgallant Flagstaff Stay	0	0	0	0	23	23
Main Topgallant Halyards	0	0	0	0	36	58
Crossjack Braces	14	21	22	30	42	50
Mizzen Topsail Pendants of Braces	2	3	3	4	4	$4\frac{1}{2}$
Mizzen Topsail Clewlines	28	31	34	38	42	54
Mizzen Topmast Tie	0	3	5	5	5	6
Mizzen Topmast Halyards	12	14	18	22	26	32
Mizzen Topsail Lifts	10	17	22	26	30	35
Mizzen Topmast Parrel Rope	0	2	3	4	5	6
Catharpin Legs and Falls	24	27	31	34	39	42
Total Weight (cwt qrs lbs)	1-0-14	2-3-17	3-3-22	5-3-24	9-0-17	9-3-12

	6th Rate	5th Rate	4th Rate	3rd Rate	2nd Rate	1st Rate
Size (ins)	1	$1\frac{1}{8}$	$1\frac{1}{4}$	$1\frac{3}{8}$	$1\frac{1}{2}$	$1\frac{3}{4}$
Spritsail Topsail Falls for Braces	16	26	51	38	41	42
Spritsail Top Halyards	5	5	8	8	$8\frac{1}{2}$	13
Spritsail Top Clewlines	14	16	18	28	$28\frac{1}{2}$	34
Spritsail Top Falls of Backstays	0	9	9	12	17	26
Spritsail Top Lifts	8	10	12	19	20	28
Fore Topgallant Mast Shrouds	8	10	13	13	14	20
Fore Topgallant Mast Halyards	24	27	28	29	30	38
Fore Topgallant Mast Lifts	17	18	18	20	25	28
Fore Topgallant Mast Braces	28	32	34	35	36	50
Fore Topgallant Mast Clewlines	28	44	50	51	52	60
Main Topgallant Mast Lanyards of the Shrouds	4	5	5	6	8	18
Main Topgallant Lanyard of the Stay	2	3	4	5	5	6
Main Topgallant Braces	46	48	50	56	58	60
Main Topgallant Mast Bowlines	42	50	53	55	56	74
Main Topgallant Mast Bridles	4	$5\frac{1}{2}$	7	7	8	11
Main Topgallant Mast Clewlines	36	46	48	48	48	49
Main Topgallant Mast Lifts	16	18	19	21	24	29
Mizzen Topmast Lanyards of Shrouds	6	8	8	10	15	20
Mizzen Topmast Braces	24	28	30	34	36	40
Mizzen Topmast Bowlines	24	27	33	35	38	44
Rattling for Shrouds	400	500	600	800	1100	1200
Total Weight (cwt qrs lbs)	3-0-4	5-0-4	5-3-19	8-3-17	9-3-21	11-0-5
Size (ins)	$0\frac{7}{8}$	1	$1\frac{1}{8}$	$1\frac{1}{4}$	$1\frac{3}{8}$	$1\frac{1}{2}$
Spritsail Topmast Lanyards of Shrouds	4	6	8	10	12	13
Fore Topgallant Lanyard of Shrouds	3	5	5	6	$7\frac{1}{2}$	14
Fore Topgallant Bowlines	32	44	52	52	60	62
Mizzen Topmast Bridles	2	4	5	5	6	9
Robins and Earrings for Sails & Clew Seizings	400	500	670	700	850	1100
Total Weight (cwt qrs lbs)	1-2-13	2-3-9	3-2-7	3-3-15	4-2-8	6-0-19
Size (ins)	$6\frac{1}{2}$	8	11	12	14	$14\frac{1}{2}$
Fore Stay	8	10	11	13	14	18
Total Weight (cwt qrs lbs)	0-2-6	1-1-0	2-2-6	3-1-10	5-1-0	7-1-16
Size (ins)	6	7	7	9	9	10
Collar of the Same	5	3	3	3	4	5
Total Weight (cwt qrs lbs)	0-0-24	0-1-3	0-1-14	0-2-0	0-2-12	0-1-1
Size (ins)	$7\frac{1}{2}$	$11\frac{1}{2}$	$13\frac{1}{2}$	14	16	17
Main Stay	11	13	14	18	$18\frac{1}{2}$	20
Total Weight (cwt qrs lbs)	1-1-4	3-1-16	4-3-25	6-3-0	9-0-9	11-1-21
Size (ins)	6	8	10	11	$12\frac{1}{2}$	15
Collar of the Same	4	5	7	8	9	$10\frac{1}{2}$
Total Weight (cwt qrs lbs)	0-1-5	0-2-14	1-1-16	1-3-16	2-3-4	3-0-24

The names, sizes, and lengths for rigging a 6th Rate demonstrated.

	Size (ins)	Fathoms
p5 **BOWSPRIT**		
Sheets	$2\frac{1}{2}$	18
Clewlines	$1\frac{1}{2}$	18
Lifts	2	12
Pendants for Braces	2	3
Falls for the Same	$1\frac{1}{2}$	28
Halyards	2	12
Horse	$2\frac{1}{3}$	$4\frac{1}{2}$
SPRITSAIL TOPMAST		
Shrouds	$1\frac{1}{2}$	8
Lanyards	$0\frac{7}{8}$	4
Pendants for Braces	$1\frac{1}{4}$	2
Falls for them	1	16
Tie	$1\frac{1}{2}$	2
Halyards	$1\frac{1}{2}$	5
Clewlines	1	14
Lifts	1	8
Parrel Ropes	1	1
FOREMAST		
Pendants for Tackles	4	4
Runners for Tackles	$2\frac{1}{2}$	8
Falls for them	$2\frac{1}{2}$	44
Shrouds	$3\frac{1}{2}$	56
Lanyards	2	26
Stay	$6\frac{1}{2}$	8
Lanyards to the Stay	2	4
Tie	0	0
Halyards	0	0
Jeers	$2\frac{5}{8}$	20
Lifts	2	34
Parrel Ropes	$2\frac{1}{3}$	7
Sheets	$2\frac{1}{2}$	36
Tacks	$3\frac{3}{4}$	18
Bowlines	2	23
Bridles	2	4
Pendants of Braces	2	4
Falls to them	$1\frac{1}{2}$	26
Clew Garnets	$1\frac{1}{2}$	34
Buntlines	$1\frac{1}{2}$	34
Collar of the Stay	6	3
FORE TOPMAST		
Shrouds	$2\frac{1}{3}$	30
Lanyards	$1\frac{1}{8}$	7
Futtocks	$1\frac{1}{2}$	6
Stay	$2\frac{1}{2}$	10
Lanyard	2	6
Standing Backstay	$2\frac{1}{2}$	21
Lanyards	$1\frac{1}{8}$	3
Lifts	$1\frac{1}{8}$	28
Tie	$2\frac{5}{8}$	5
The Fall of the Halyards	2	18
Bowlines	$1\frac{1}{2}$	40

	Size (ins)	Fathoms
Bridles	0	0
Pendants for Braces	$1\frac{1}{2}$	3
Braces	$1\frac{1}{2}$	30
Sheets	$2\frac{5}{8}$	21
Clewlines	$1\frac{1}{2}$	40
Parrel Rope	$1\frac{1}{2}$	3
Buntlines	0	0
FORE TOPGALLANT MAST		
Shrouds	1	8
Lanyards	0	0
Top Rope	0	0
Stay	$1\frac{1}{8}$	11
Lanyard	0	0
Tie	$1\frac{1}{2}$	2
Halyards	1	24
Lifts	1	17
Pendants of Braces	0	0
Clewlines	1	28
Bowlines and Bridles	0	0
Braces	1	28
MAINMAST		
Pendants of Tackles	$4\frac{1}{4}$	3
Runners of Tackles	0	0
Falls of Tackles	$2\frac{1}{8}$	50
Shrouds	$4\frac{1}{4}$	76
Lanyards	$2\frac{1}{8}$	30
Standing Backstay	0	0
Lanyards	0	0
Stay	$7\frac{1}{2}$	11
Collar of the Stay	6	4
Lanyard of the Stay	$2\frac{1}{3}$	6
Lifts	2	42
Tacks	4	22
Sheets	3	40
Bowlines	$2\frac{1}{2}$	22
Bridles	0	0
Pendants	$2\frac{1}{2}$	5
Braces	2	40
Clew Garnets	2	41
Jeers	$3\frac{3}{4}$	39
Parrel Ropes	$2\frac{1}{3}$	8
Breast Ropes	0	0
Buntlines	$1\frac{1}{2}$	38
Pendant of the Garnet	4	5
Guy	$2\frac{1}{3}$	7
Fall of the Garnet	3	24
MAIN TOPMAST		
Pendant of Tackles	2	4
Falls of Tackles	$1\frac{1}{2}$	18
Shrouds	$2\frac{1}{2}$	34
Lanyards	$1\frac{1}{8}$	14
Standing Backstays	$2\frac{1}{3}$	26
Lanyards	0	0

	Size (ins)	Fathoms
Stay	$2\frac{1}{2}$	14
Lanyard	2	4
Lifts	$1\frac{1}{2}$	40
Pendants	2	3
Braces	$1\frac{1}{2}$	42
Bowlines	$2\frac{1}{8}$	36
Bridles	2	6
Sheets	$3\frac{1}{2}$	28
Clewlines	2	44
Tie	$3\frac{3}{4}$	6
Runners of the Halyards	$2\frac{1}{2}$	10
Fall of the Halyards	2	36
Parrel Ropes	2	5
Buntlines	$1\frac{1}{2}$	18
Futtocks	3	16

MAIN TOPGALLANT MAST

	Size (ins)	Fathoms
Shrouds	$1\frac{1}{8}$	10
Lanyards	0	0
Futtocks	$1\frac{1}{8}$	4
Stay	$1\frac{1}{8}$	8
Lanyard of the Stay	0	0
Pendants	2	3
Braces	1	46
Bowlines	0	0
Bridles	0	0

	Size (ins)	Fathoms
Parrel Ropes	0	0
Tie	$1\frac{1}{2}$	2
Halyards	1	28
Clewlines	1	36
Lifts	1	16

MIZZEN MAST

	Size (ins)	Fathoms
Shrouds	$2\frac{1}{2}$	36
Lanyards	$1\frac{1}{2}$	10
Tie	0	0
Halyards	0	0
Stay	$2\frac{5}{8}$	$7\frac{1}{2}$
Lanyard	$1\frac{1}{2}$	2
Sheet	$2\frac{1}{8}$	14
Jeer	$2\frac{1}{2}$	20
Truss	0	0
Bowlines	2	7
Brails	1	40
Parrel Ropes	$2\frac{1}{8}$	4

CROSSJACK

	Size (ins)	Fathoms
Lifts	2	5
Braces	$1\frac{1}{8}$	14
Pendants	0	0
Halyards	0	0
Slings	$2\frac{1}{8}$	3

MIZZEN TOPMAST

	Size (ins)	Fathoms
Shrouds	$1\frac{1}{2}$	12
Lanyards	1	6
Futtocks	$1\frac{1}{2}$	3
Pendants	$1\frac{1}{8}$	2
Braces	1	24
Clewlines	$1\frac{1}{8}$	28
Tie	0	0
Halyards	0	0
Lifts	$1\frac{1}{8}$	10
Parrel Ropes	0	0
Falls	0	0
Stoppers at the Bow	$3\frac{1}{2}$	14
Shank Painter	$3\frac{3}{4}$	12
Slings for Ordnance	6	4
Boat Rope	$4\frac{1}{2}$	26
2 Pair of Butt Slings	5	5
2 Pair of Hogshead Slings	4	4
For the Bowsprit Wolding	4	34
For the Stem Wolding	4	34
Stoppers at the Bitts	$4\frac{1}{2}$	4
Lanyards	$2\frac{1}{8}$	6
For Robins, Earrings and Clew Seizings for one Complete Suit of Sails	$1\frac{1}{2}$	100
	1	175

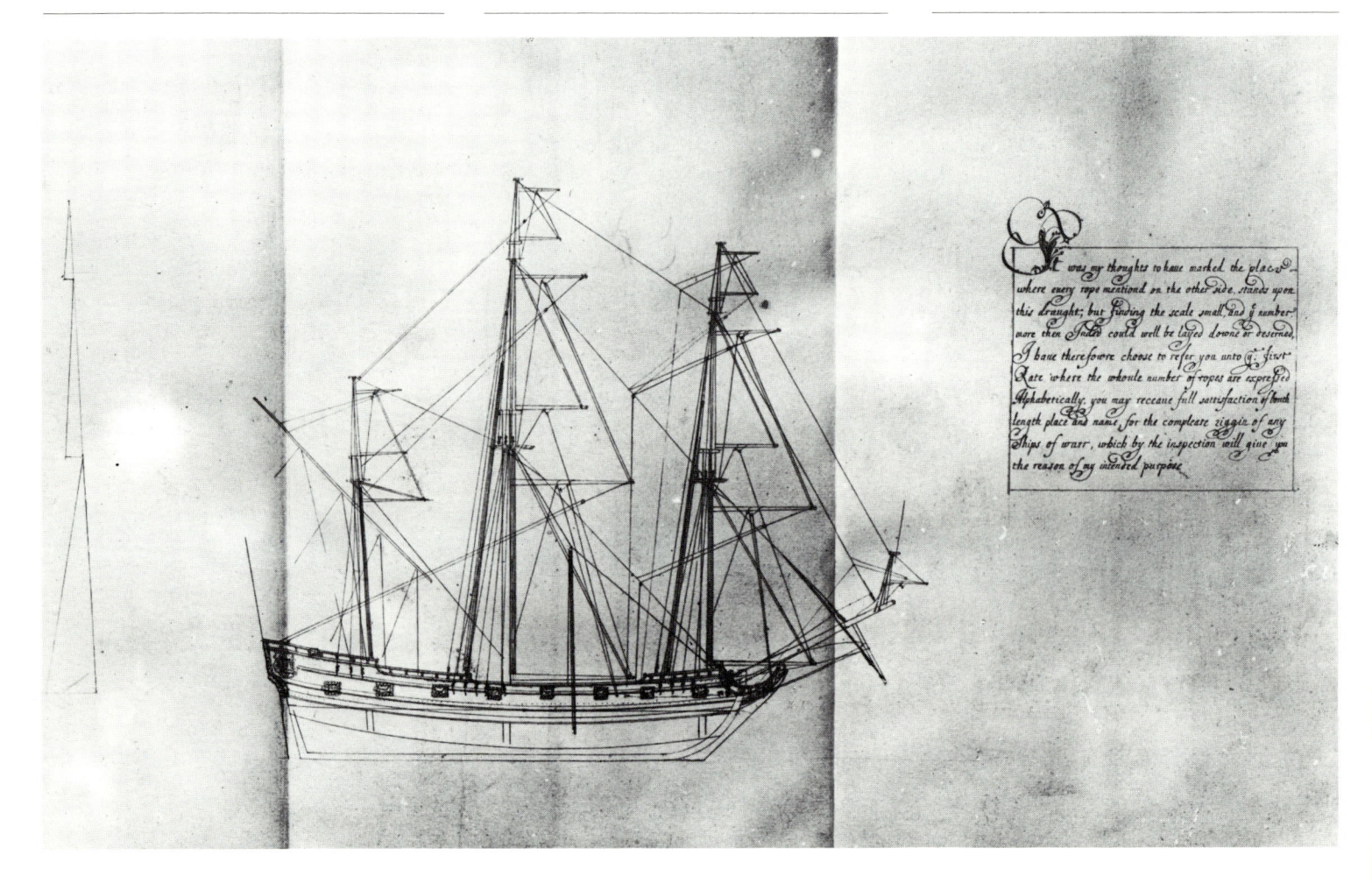

The names, sizes, and lengths for rigging a 5th Rate demonstrated.

	Size (ins)	Fathoms
BOWSPRIT		
Sheets	$2\frac{7}{8}$	50
Clewlines	$1\frac{9}{10}$	24
Garnets	$1\frac{9}{10}$	24
Lifts	$2\frac{1}{2}$	16
Pendants for Braces	$2\frac{7}{8}$	4
Falls for the Same	$1\frac{9}{10}$	30
Halyards	$2\frac{1}{2}$	20
Tie	0	0
Buntlines	$1\frac{1}{2}$	20
Horse	$3\frac{1}{3}$	6
Slings	0	0
SPRITSAIL TOPMAST		
Shrouds	$1\frac{9}{10}$	12
Lanyards	1	6
Pendants for Braces	$1\frac{1}{2}$	3
Falls for them	$1\frac{1}{8}$	26
Tie	$1\frac{9}{10}$	2
Halyards	$1\frac{1}{8}$	5
Clewlines	$1\frac{1}{8}$	16
Pendants for Backstays	$1\frac{1}{2}$	3
Falls for the Same	$1\frac{1}{8}$	9
Lifts	$1\frac{1}{8}$	10
Parrel Ropes	$1\frac{1}{2}$	1
FOREMAST		
Pendants for Tackles	5	5
Runners for Tackles	$4\frac{3}{8}$	21
Falls for them	$2\frac{1}{2}$	46
Shrouds	5	80
Lanyards	$2\frac{7}{8}$	30
Standing Backstays	0	0
Lanyard	0	0
Stay	8	10
Lanyards to the Stay	3	7
Tie	0	0
Halyards	0	0
Jeers	$4\frac{3}{8}$	40
Lifts	$2\frac{1}{2}$	46
Parrel Ropes	3	7
Breast Ropes	0	0
Sheets	4	48

	Size (ins)	Fathoms
Tacks	$4\frac{3}{4}$	27
Bowlines	3	27
Bridles	2	4
Pendants	$2\frac{7}{8}$	4
Braces	$2\frac{1}{2}$	34
Clew Garnets	$1\frac{9}{10}$	44
Buntlines	$1\frac{1}{2}$	46
Collar of the Stay	7	6
FORE TOPMAST		
Shrouds	3	42
Lanyards	$2\frac{1}{2}$	8
Futtocks	$1\frac{9}{10}$	20
Stay	4	11
Pendant of the Lanyard	0	0
Fall of the Lanyard	0	0
Tie	$4\frac{3}{8}$	$6\frac{1}{2}$
Runner of the Halyards	$3\frac{1}{2}$	$8\frac{1}{2}$
Fall to them	$2\frac{1}{2}$	34
Standing Backstays	$3\frac{1}{4}$	55
Lanyards to them	$1\frac{1}{2}$	6
Lifts	$1\frac{1}{2}$	48
Bowlines	$1\frac{9}{10}$	48
Bridles	$1\frac{1}{2}$	10
Pendants for Braces	$2\frac{1}{2}$	4
Falls to them	$1\frac{9}{10}$	40
Pendant of the Top Rope	0	0
Fall for the Same	0	0
Sheets	$4\frac{3}{8}$	34
Clewlines	$1\frac{9}{10}$	56
Parrel Ropes	$2\frac{1}{2}$	5
Buntlines	$1\frac{1}{2}$	28
FORE TOPGALLANT MAST		
Shrouds	11	10
Lanyards	1	5
Futtocks	0	0
Lanyard	0	0
Tie	$1\frac{9}{10}$	2
Halyards	$1\frac{1}{8}$	27
Lifts	$1\frac{1}{8}$	18
Pendants	0	0
Braces	$1\frac{1}{8}$	32
Clewlines	$1\frac{1}{8}$	44
Bowlines and Bridles	1	44
MAINMAST		
Pendants of Tackles	$5\frac{3}{8}$	4
Runners of Tackles	$4\frac{3}{8}$	24
Falls of Tackles	$2\frac{7}{8}$	56
Shrouds	$5\frac{1}{2}$	114
Lanyards	$2\frac{7}{8}$	40
Standing Backstays	0	0
Lanyards	0	0
Stay	$11\frac{1}{2}$	13
Collar of the Stay	8	5
Lanyards of the Stay	$3\frac{1}{3}$	10

	Size (ins)	Fathoms
Lifts	$2\frac{1}{2}$	44
Tacks	5	26
Sheets	4	50
Bowlines	$2\frac{7}{8}$	28
Bridles	3	10
Pendants	$2\frac{7}{8}$	5
Braces	$2\frac{1}{2}$	44
Clew Garnets	$2\frac{1}{2}$	44
Tie	0	0
Halyards	0	0
Jeers	$4\frac{3}{4}$	60
Parrel Ropes	$3\frac{1}{3}$	10
Buntlines	$1\frac{9}{10}$	50
Pendant of the Garnet	$4\frac{1}{2}$	8
Guy	$3\frac{1}{3}$	8
Fall of the Garnet	$3\frac{1}{4}$	25
Breast Ropes	$4\frac{3}{8}$	3
MAIN TOPMAST		
Pendants of Tackles	$2\frac{7}{8}$	5
Falls of Tackles	$1\frac{1}{2}$	28
Shrouds	$3\frac{1}{4}$	48
Lanyards	$1\frac{1}{2}$	15
Standing Backstays	$3\frac{1}{3}$	64
Lanyards	0	0
Stay	$3\frac{1}{2}$	16
Lanyards	2	8
Lifts	$1\frac{9}{10}$	54
Pendants	2	4
Braces	$1\frac{9}{10}$	48
Bowlines	$2\frac{7}{8}$	40
Bridles	$2\frac{1}{2}$	9
Pendant of the Top Rope	$4\frac{1}{2}$	9
Fall of the Same	$3\frac{1}{3}$	36
Sheets	5	42
Clewlines	$2\frac{1}{2}$	60
Tie	$4\frac{3}{4}$	6
Runner of the Halyards	4	11
Fall of the Halyards	$2\frac{1}{2}$	46
Parrel Ropes	$2\frac{1}{2}$	5
Buntlines	$1\frac{9}{10}$	30
Futtocks	$3\frac{1}{2}$	20
MAIN TOPGALLANT MAST		
Shrouds	$1\frac{1}{2}$	12
Lanyards	$1\frac{1}{8}$	5
Futtocks	$1\frac{1}{2}$	12
Stay	$1\frac{1}{2}$	16
Lanyard of the Stay	$1\frac{1}{8}$	3
Pendants	$1\frac{1}{2}$	0
Braces	$1\frac{1}{8}$	48
Bowlines	$1\frac{1}{8}$	50
Bridles	$1\frac{1}{8}$	$5\frac{1}{2}$
Parrel Ropes	$1\frac{1}{2}$	1
Tie	$1\frac{9}{10}$	3
Halyards	$1\frac{1}{2}$	30
Clewlines	$1\frac{1}{8}$	46
Lifts	$1\frac{1}{8}$	18

	Size (ins)	Fathoms
MIZZEN MAST		
Pendant of the Tackles	0	0
Runners	0	0
Falls of the Tackles	0	0
Shrouds	$3\frac{1}{8}$	41
Lanyards	$1\frac{9}{10}$	8
Tie	0	0
Halyards	0	0
Stay	$4\frac{3}{8}$	9
Lanyard	$1\frac{9}{10}$	2
Sheet	$2\frac{7}{8}$	15
Jeer	$3\frac{1}{4}$	22
Truss	$1\frac{9}{10}$	12
Bowlines	2	9
Brails	$1\frac{9}{10}$	44
Parrel Ropes	3	6
CROSSJACK		
Lifts	$2\frac{1}{4}$	6
Braces	$1\frac{1}{2}$	21
Pendants	0	0
Halyards	0	0
Slings	$2\frac{7}{8}$	4

	Size (ins)	Fathoms
MIZZEN TOPMAST		
Shrouds	$1\frac{9}{10}$	14
Lanyards	$0\frac{3}{4}$	8
Futtocks	$1\frac{9}{10}$	9
Pendants	$1\frac{1}{2}$	3
Braces	1	28
Bowlines	1	27
Bridles	1	4
Clewlines	$1\frac{1}{2}$	31
Tie	$1\frac{1}{2}$	3
Halyards	$1\frac{1}{2}$	14
Lifts	1	17
Top Rope	0	0
Parrel Ropes	$1\frac{1}{2}$	2
Cat Rope/Pendants	4	40
Cat Rope/Falls	0	0
Pendant of the Fish Hook Rope	5	6
Fall of the Same	3	26
Stoppers at the Bow	4	14
Shank Painter	$4\frac{3}{4}$	15
4 Buoy Ropes Cables	$5\frac{3}{8}$	50
Buoy Rope for Stream Anchor	$2\frac{1}{3}$	15

	Size (ins)	Fathoms
Buoy Rope for Kedge Anchor	3	10
Slings for Ordnance	6	4
Viol Cable	0	0
Passing Rope	0	0
Boat Rope Cables	5	30
Pinnace Rope Cables	$3\frac{1}{3}$	26
Gust Rope Cables	$3\frac{1}{3}$	26
Pair of Butt Slings	5	6
Pair of Hogshead Slings	4	4
Wolding for the Bowsprit	$4\frac{1}{2}$	50
Wolding for the Stem	$4\frac{1}{2}$	50
Stoppers at the Bitts	5	6
Lanyards	$2\frac{7}{8}$	7
For Robins, Earrings, and Clew Seizings for one Suit of Sails	$1\frac{1}{2}$ / 1	400 / 700

The names, sizes, and lengths for rigging a 4th Rate demonstrated.

	Size (ins)	Fathoms
BOWSPRIT		
Pendants for Sheets	0	0
Falls for the Same	$3\frac{1}{8}$	54
Clewlines	$2\frac{1}{8}$	30
Garnets	$2\frac{1}{8}$	40
Lifts	$2\frac{3}{4}$	18
Pendants for Braces	$2\frac{3}{4}$	4
Falls for the Same	$2\frac{3}{8}$	48
Halyards	$2\frac{3}{8}$	22
Tie	0	0
Buntlines	2	25
Horse	$3\frac{3}{4}$	$6\frac{1}{2}$
Slings	$3\frac{3}{4}$	$3\frac{1}{2}$
SPRITSAIL TOPMAST		
Shrouds	$2\frac{1}{8}$	14
Lanyards	$1\frac{1}{8}$	8
Pendants	2	3
Braces	1	34
Tie	$2\frac{1}{8}$	2
Halyards	$1\frac{1}{4}$	8
Clewlines	$1\frac{3}{8}$	18
Pendants for Backstays	2	3
Falls for the Same	$1\frac{1}{4}$	9
Lifts	$1\frac{1}{4}$	12
Parrel Ropes	2	1
FOREMAST		
Pendants for Tackles	6	$5\frac{1}{2}$
Runners for Tackles	$5\frac{1}{8}$	32
Falls for them	$2\frac{3}{4}$	52
Shrouds	6	118
Lanyards	$3\frac{1}{2}$	36
Standing Backstays	0	0
Lanyard	0	0
Stay	11	11
Lanyards to the Stay	3	8
Tie	0	0
Halyards	0	0
Jeers	$5\frac{1}{8}$	70
Lifts	$2\frac{3}{4}$	48
Parrel Ropes	$3\frac{1}{4}$	$7\frac{1}{2}$
Breast Ropes	0	0
Sheets	$4\frac{1}{2}$	54

	Size (ins)	Fathoms
Tacks	$5\frac{3}{4}$	28
Bowlines	$3\frac{1}{4}$	32
Bridles	3	4
Pendants	$2\frac{3}{4}$	$4\frac{1}{2}$
Braces	$2\frac{3}{4}$	46
Clew Garnets	$2\frac{1}{8}$	46
Buntlines	2	60
Collar of the Stay	7	3
FORE TOPMAST		
Shrouds	$3\frac{1}{4}$	58
Lanyards	2	12
Futtocks	$2\frac{1}{8}$	24
Stay	$4\frac{1}{2}$	15
Pendant of the Lanyard	$2\frac{1}{3}$	10
Fall of the Lanyard	0	0
Tie	$5\frac{1}{8}$	7
Runner of the Halyards	$3\frac{1}{2}$	$9\frac{1}{2}$
The Fall to them	$2\frac{1}{2}$	35
Standing Backstays	$3\frac{1}{2}$	59
Lanyards to them	2	8
Lifts	2	50
Bowlines	$2\frac{1}{8}$	50
Bridles	2	10
Pendants for Braces	$2\frac{1}{2}$	4
Falls to them	$2\frac{1}{8}$	49
Pendant for the Top Rope	$5\frac{3}{4}$	12
Fall of the Same	$3\frac{1}{2}$	34
Sheets	$5\frac{1}{8}$	36
Clewlines	$2\frac{1}{8}$	64
Parrel Rope	$2\frac{1}{8}$	6
Buntlines	2	30
FORE TOPGALLANT MAST		
Parrel Ropes	2	$1\frac{1}{2}$
Shrouds	$1\frac{1}{4}$	13
Lanyards	$1\frac{1}{8}$	5
Futtocks	0	0
Lanyard	0	0
Tie	$2\frac{1}{8}$	$2\frac{1}{2}$
Halyards	$1\frac{1}{4}$	28
Lifts	$1\frac{1}{4}$	18
Pendants	0	0
Braces	$1\frac{1}{4}$	34
Clewlines	$1\frac{1}{4}$	50
Bowlines and Bridles	$1\frac{1}{8}$	52
Stay	2	$15\frac{1}{2}$
MAINMAST		
Pendants of Tackles	$6\frac{1}{4}$	4
Runners of Tackles	$5\frac{1}{8}$	26
Shrouds	$6\frac{1}{4}$	152
Lanyards	$3\frac{1}{4}$	46
Standing Backstays	0	0
Lanyards	0	0
Stay	$13\frac{1}{2}$	14
Collar of the Stay	10	7

	Size (ins)	Fathoms
Lifts	$2\frac{3}{4}$	54
Tacks	$6\frac{1}{2}$	30
Sheets	$5\frac{1}{2}$	60
Bowlines	$3\frac{1}{4}$	34
Bridles	$2\frac{1}{2}$	10
Pendants	$3\frac{1}{8}$	$5\frac{1}{2}$
Braces	$2\frac{3}{4}$	60
Clew Garnets	$2\frac{3}{4}$	46
Tie	0	0
Halyards	0	0
Jeers	$5\frac{3}{4}$	66
Parrel Ropes	$3\frac{3}{4}$	12
Buntlines	$2\frac{1}{8}$	58
Pendant of the Garnet	$5\frac{1}{2}$	9
Guy	$3\frac{3}{4}$	8
Fall of the Garnet	$3\frac{1}{8}$	27
Lanyard of the Stay	$3\frac{3}{4}$	12
MAIN TOPMAST		
Pendants of Tackles	$3\frac{1}{8}$	6
Falls of Tackles	2	29
Shrouds	$3\frac{1}{2}$	58
Lanyards	2	16
Standing Backstays	$3\frac{3}{4}$	75
Lanyards	$2\frac{1}{4}$	14
Stay	5	18
Lanyard	$2\frac{1}{2}$	8
Lifts	$2\frac{1}{8}$	60
Pendants for Braces	$2\frac{1}{2}$	5
Falls to them	$2\frac{1}{8}$	60
Bowlines	$3\frac{1}{8}$	48
Bridles	$2\frac{1}{2}$	10
Pendant of the Top Rope	$5\frac{1}{2}$	9
Fall of the Same	$3\frac{3}{4}$	38
Sheets	6	44
Clewlines	$2\frac{3}{4}$	74
Tie	$5\frac{3}{4}$	$7\frac{1}{2}$
Runner of the Halyards	$4\frac{1}{2}$	12
Fall of the Halyards	$2\frac{3}{4}$	50
Parrel Ropes	$2\frac{3}{4}$	6
Buntlines	$2\frac{1}{8}$	32
Futtocks	$3\frac{3}{4}$	22
MAIN TOPGALLANT MAST		
Shrouds	2	14
Lanyards	$1\frac{1}{4}$	5
Stay	$1\frac{3}{4}$	17
Lanyard of the Stay	$1\frac{1}{4}$	4
Pendants for Braces	0	0
Braces	$1\frac{1}{4}$	50
Bowlines	$1\frac{1}{4}$	53
Bridles	$1\frac{1}{4}$	7
Parrel Ropes	2	2
Tie	$2\frac{1}{8}$	3
Halyards	2	31
Clewlines	$1\frac{1}{4}$	48
Lifts	$1\frac{3}{8}$	19

	Size (ins)	Fathoms
MIZZEN MAST		
Pendants of Tackles	0	0
Runners	0	0
Falls of Tackles	0	0
Shrouds	$3\frac{1}{2}$	62
Lanyards	$2\frac{1}{8}$	14
Tie	0	0
Halyards	0	0
Stay	$5\frac{5}{8}$	11
Lanyard	$2\frac{1}{8}$	$2\frac{1}{2}$
Sheet	$3\frac{3}{8}$	16
Jeer	$3\frac{1}{2}$	30
Truss	$2\frac{1}{8}$	16
Bowlines	$2\frac{1}{2}$	9
Bridles	$2\frac{1}{8}$	68
Parrel Ropes	$3\frac{1}{2}$	6
CROSSJACK		
Lifts	$2\frac{3}{4}$	7
Braces	2	22
Pendants	0	0
Halyards	0	0
Slings	$3\frac{1}{8}$	5

	Size (ins)	Fathoms
MIZZEN TOPMAST		
Falls of Tackles	0	0
Shrouds	$2\frac{1}{8}$	20
Lanyards	1	8
Futtocks	$2\frac{1}{8}$	9
Pendants	$1\frac{1}{2}$	3
Braces	1	30
Bowlines	$1\frac{1}{2}$	33
Bridles	1	5
Clewlines	$1\frac{1}{2}$	34
Tie	2	5
Halyards	$1\frac{3}{4}$	18
Lifts	$1\frac{1}{4}$	22
Top Rope	0	0
Parrel Rope	$1\frac{1}{2}$	3
Falls	0	0
Cat Rope Pendants and Falls	$4\frac{1}{2}$	43
Pendant of the Fish Hook Rope	6	7
Fall of the Same	$3\frac{1}{4}$	28
Stoppers at the Bow	7	14
Shank Painter	$5\frac{3}{4}$	15

	Size (ins)	Fathoms
Buoy Ropes–Cables of	$6\frac{1}{4}$	50
Buoy Rope for the Stream Anchor	$3\frac{3}{4}$	15
Slings for Ordnance	6	$4\frac{1}{2}$
Viol Cable	0	0
Passing Rope	0	0
Boat Rope Cables	$5\frac{1}{2}$	30
Pinnace Rope Cables	$3\frac{3}{4}$	28
Gust Rope Cables	$3\frac{3}{4}$	28
2 Pair of Butt Slings	5	6
2 Pair of Hogshead Slings	4	4
Wolding for the Bowsprit and Stem	$4\frac{3}{4}$	60
Stoppers at the Bitts	6	7
Lanyards	$3\frac{1}{8}$	8
For Robins, Earrings and	$1\frac{1}{2}$	400
Clew Siezings for one Complete Suit of Sails	1	800

The names, sizes, and lengths for rigging a 3rd Rate demonstrated.

BOWSPRIT

	Size (ins)	Fathoms
Sheets	$3\frac{3}{5}$	58
Horse	$4\frac{1}{4}$	$7\frac{1}{2}$
Clewlines	$2\frac{3}{8}$	34
Slings for the Yard	$4\frac{1}{4}$	4
Halyards	3	22
Lifts	$3\frac{1}{10}$	42
Standing Lifts	0	0
Lanyards	0	0
Pendants of Braces	$2\frac{3}{4}$	4
Falls	$2\frac{3}{8}$	46
Wolding	5	68
Garnets	$2\frac{3}{8}$	48
Buntlines	$2\frac{1}{8}$	28

SPRITSAIL TOPMAST

	Size (ins)	Fathoms
Shrouds	$2\frac{3}{8}$	16
Lanyards	$1\frac{1}{4}$	10
Pendants of Braces	$2\frac{1}{6}$	3
Falls	$1\frac{1}{2}$	38
Tie	$2\frac{3}{8}$	3
Halyards	$1\frac{3}{8}$	8
Clewlines	$1\frac{3}{8}$	28
Pendants of Backstays	$2\frac{1}{8}$	4
Falls	$1\frac{3}{8}$	12
Lifts	$1\frac{3}{8}$	19
Parrel Ropes	$2\frac{1}{8}$	2

FOREMAST

	Size (ins)	Fathoms
Pendants of Tackles	$6\frac{3}{4}$	$6\frac{1}{2}$
Runners of Tackles	$5\frac{1}{2}$	25
Falls of Tackles	$3\frac{1}{10}$	60
Shrouds	$6\frac{3}{4}$	136
Lanyards	$3\frac{1}{2}$	56
Stay	12	13
Collar	8	3
Lanyard	$3\frac{3}{4}$	10
Futtocks	0	0
Parrel Ropes	$3\frac{3}{4}$	12
Clew Garnets	$2\frac{3}{8}$	52
Bowlines	$3\frac{3}{4}$	46
Bridles	3	5
Pendants of Braces	$3\frac{1}{10}$	5
Falls	$3\frac{1}{10}$	50
Buntlines	$2\frac{1}{8}$	62
Sheets	$4\frac{7}{8}$	60
Tacks	$6\frac{1}{3}$	32
Lifts	$3\frac{1}{10}$	54
Leech Lines	2	48
Jeers	$5\frac{1}{2}$	82
Legs for Catharpins	2	14
Falls	2	12
Stoppers for Topsail Sheets	0	0
Tie	0	0
Halyards	0	0

FORE TOPMAST

	Size (ins)	Fathoms
Pendant of the Top Rope	$6\frac{1}{3}$	13
Fall of the Same	$4\frac{1}{8}$	36
Shrouds	$3\frac{3}{4}$	64
Lanyards	$2\frac{1}{8}$	19
Standing Backstays	$4\frac{1}{8}$	66
Lanyards	$2\frac{1}{8}$	9
Stay	$4\frac{7}{8}$	$15\frac{1}{2}$
Lanyard	$3\frac{1}{10}$	10
Lifts	$2\frac{1}{8}$	54
Futtocks	$2\frac{3}{8}$	30
Tie	$5\frac{1}{2}$	7
Runner	$4\frac{1}{8}$	14
Halyards	3	39
Bowlines	$2\frac{3}{8}$	58
Bridles	$2\frac{1}{8}$	10
Clewlines	$2\frac{3}{8}$	66
Pendants of Tackles	$2\frac{1}{2}$	$4\frac{1}{2}$
Falls	2	28
Sheets	$5\frac{1}{2}$	44
Parrel Ropes	$2\frac{3}{8}$	7
Leech Lines	$1\frac{3}{4}$	12
Braces	$2\frac{3}{8}$	54
Pendants	$6\frac{1}{3}$	14
Buntlines	$2\frac{1}{8}$	32

FORE TOPGALLANT MAST

	Size (ins)	Fathoms
Stay	$2\frac{1}{8}$	16
Tie	$2\frac{3}{8}$	3
Halyards	$1\frac{3}{8}$	29
Lifts	$1\frac{3}{8}$	20
Braces	$1\frac{3}{8}$	35
Clewlines	$1\frac{3}{8}$	51
Bowlines and Bridles	$1\frac{1}{4}$	52
Parrel Ropes	$2\frac{1}{8}$	$1\frac{1}{2}$
Shrouds	$1\frac{3}{8}$	13
Lanyards	1	6

MAINMAST

	Size (ins)	Fathoms
Pendants of Tackles	7	5
Runners of Tackles	$5\frac{1}{2}$	28
Falls of Tackles	$3\frac{2}{5}$	62
Lifts	$3\frac{1}{10}$	66
Shrouds	7	182
Lanyards	0	0
Stay	14	18
Collar	11	8
Lanyard	$4\frac{1}{4}$	13
Pendant of the Garnet	6	10
Guy	$4\frac{1}{4}$	9
Fall of the Garnet	$3\frac{3}{4}$	34
Tacks	8	34
Sheets	6	66
Clewgarnets	$3\frac{1}{10}$	58
Bowlines	4	38
Bridles	3	10
Pendants of Braces	$3\frac{2}{5}$	6
Falls	$3\frac{1}{10}$	64
Jeers	$6\frac{1}{3}$	96
Parrel Ropes	$4\frac{1}{4}$	14
Leech Lines	2	65
Buntlines	$2\frac{3}{8}$	78
Futtocks	0	0
Legs of Catharpins	$2\frac{1}{4}$	17
Falls	2	16
Stoppers for Topsail Sheets	0	0
2 Tackles to Hoist up Shrouds	0	0
Bowsing Tackles	3	11
Knave Line	2	18
Horse for the Yard	$4\frac{1}{2}$	13

MAIN TOPMAST

	Size (ins)	Fathoms
Pendants of Tackles	$3\frac{2}{5}$	6
Falls	$2\frac{1}{8}$	30
Shrouds	$4\frac{1}{8}$	76
Lanyards	$2\frac{1}{8}$	26
Standing Backstays	$4\frac{1}{4}$	108
Lanyards	3	12
Stay	5	21
Lanyard	$2\frac{1}{8}$	9
Lifts	$2\frac{3}{8}$	62
Braces	$2\frac{3}{8}$	64
Pendants	3	5
Bowlines	$3\frac{2}{5}$	50
Bridles	3	14
Pendant of the Top Rope	$7\frac{1}{2}$	$14\frac{1}{2}$
Fall of the Same	$4\frac{1}{4}$	42
Clewlines	$3\frac{1}{10}$	76
Tie	$6\frac{1}{3}$	9
Runner	5	17
Halyards	3	56
Leech Lines	$2\frac{1}{4}$	16
Buntlines	$2\frac{3}{8}$	44
Futtocks	4	28
Parrel Ropes	$3\frac{1}{10}$	8
Sheets	$6\frac{3}{4}$	48

DEANE'S DOCTRINE OF NAVAL ARCHITECTURE, 1670

	Size (ins)	Fathoms
MAIN TOPGALLANT MAST		
Stay	2	$19\frac{1}{2}$
Braces	$1\frac{3}{8}$	56
Bowlines	$1\frac{3}{8}$	55
Bridles	$1\frac{3}{8}$	$7\frac{1}{2}$
Parrel Ropes	$2\frac{1}{8}$	3
Tie	$2\frac{1}{8}$	3
Halyards	$2\frac{1}{8}$	35
Clewlines	$1\frac{3}{8}$	48
Lifts	$1\frac{3}{8}$	21
Shrouds	$2\frac{1}{8}$	15
Lanyards	$1\frac{3}{8}$	6
MIZZEN MAST		
Burtons	0	0
Falls of Burtons	0	0
Shrouds	4	84
Lanyards	4	84
Stay	$5\frac{1}{2}$	13
Lanyard	$2\frac{3}{8}$	3
Halyards	0	0
Parrel Ropes	$3\frac{1}{2}$	6
Truss	$2\frac{3}{8}$	26
Sheet	$3\frac{2}{5}$	18
Tack	0	0
Bowlines	3	10

	Size (ins)	Fathoms
Brails	$2\frac{3}{8}$	80
Jeers	$4\frac{1}{8}$	32
CROSSJACK		
Lifts	0	0
Braces	$2\frac{1}{8}$	30
Standing Lifts	3	8
Lanyards	0	0
Slings	$3\frac{2}{5}$	5
MIZZEN TOPMAST		
Shrouds	$2\frac{3}{8}$	26
Lanyards	$1\frac{1}{2}$	10
Futtocks	$2\frac{3}{8}$	4
Braces	$1\frac{1}{2}$	34
Bowlines	$1\frac{1}{2}$	35
Bridles	1	5
Tie	$2\frac{1}{8}$	5
Halyards	2	22
Lifts	2	26
Parrel Ropes	$1\frac{3}{4}$	4
Stay	2	8
Clewlines	$1\frac{3}{4}$	38
Cat Ropes	$4\frac{7}{8}$	45
Pendant of the Fish Hook Rope	$6\frac{3}{4}$	8

	Size (ins)	Fathoms
Fall of the Same	$3\frac{3}{4}$	30
Stopper at the Bow	$7\frac{1}{2}$	14
Shank Painter	$6\frac{1}{2}$	15
Stream Anchor Stopper	$4\frac{1}{4}$	10
Shank Painter	0	0
Stopper at the Bitts	7	8
Lanyards	$3\frac{2}{5}$	10
Viol	0	0
Pendant of the Winding Tackle	10	10
Fall of the Same	5	40
Buoy Ropes	$4\frac{1}{2}$	15
Stream Anchor Buoy Ropes	$4\frac{1}{4}$	15
Boat Rope and Slings	7	32
Gust Rope to It	$4\frac{1}{4}$	30
Pinnace Rope and Slings	$4\frac{1}{4}$	30
Gust Rope to It	$4\frac{1}{2}$	30
2 Pairs of Butt Slings	5	7
2 Pairs of Hogshead Slings	4	8
Ordnance Slings	8	5
For Robins, Earrings, and Clew Siezings for one Complete Suit of Sails	$1\frac{3}{4}$	1000

The names, sizes, and lengths for rigging a 2nd Rate demonstrated.

	Size (ins)	Fathoms
p13 **BOWSPRIT**		
Sheets	3½	60
Horse	5	8
Clewlines	2½	39
Slings for the Yard	5	5
Halyards	3	27
Lifts	3	50
Standing Lifts	0	0
Lanyards	0	0
Pendants of Braces	3	4
Falls	2½	47
Wolding	6	80
Garnets	2½	55
Buntlines	2	37
SPRITSAIL TOPMAST		
Shrouds	2½	18
Lanyards	1	12
Pendants of Braces	2	3
Falls	1½	41
Tie	2½	3
Halyards	1½	8½
Clewlines	1½	28½
Pendants of Backstays	2	5½
Falls	1½	17
Lifts	1½	20
Parrel Ropes	2	3½
FOREMAST		
Pendants of Tackles	7½	7½
Runners of Tackles	6	26
Falls of Tackles	3	151
Shrouds	7½	188
Lanyards	3½	60
Stay	14	14
Collar	9	4
Lanyard	4	12
Futtock	4	24
Parrel Ropes	4	14½
Clew Garnets	2½	57
Bowlines	4	56
Bridles	3	6
Pendants of Braces	3½	6

	Size (ins)	Fathoms
Falls	3	52
Buntlines	2	68
Sheets	5½	56
Tacks	7	36
Lifts	3	58
Leech Lines	3	58
Jeers	6	86
Legs for Catharpins	2½	16
Falls	2½	17
Stoppers for Topsail Sheets	0	0
Tie	0	0
Halyards	0	0
FORE TOPMAST		
Pendant of the Top Rope	7	14
Fall of the Same	4	44
Shrouds	4	62
Lanyards	2	20
Standing Backstays	4½	114
Lanyards	2	14
Stay	5¼	16½
Lanyard	3	10
Lifts	2	58
Futtocks	2½	48
Tie	6	7½
Runner	4½	16
Halyards	3	40
Bowlines	2½	62
Bridles	2	11
Clewlines	2½	74
Pendants of Tackles	3	5½
Falls	2	34
Sheets	6	45
Parrel Ropes	2½	8
Leech Lines	2	14
Braces	2½	54
Pendants	3	4
Buntlines	2	34
FORE TOPGALLANT MAST		
Stay	2	19½
Tie	2½	3½
Halyards	1½	30
Lifts	1½	25
Braces	1½	36
Clewlines	1½	52
Bowlines	1	70
Bridles	1	8
Parrel Ropes	2	2
Shrouds	1½	14
Lanyards	1	7½
MAINMAST		
Pendants of Tackles	8	8
Runners of Tackles	6	31

	Size (ins)	Fathoms
Falls of Tackles	3½	160
Lifts	3	70
Shrouds	8¼	216
Lanyards	3½	76
Stay	16	18½
Collar	12½	9
Lanyard	5	14
Pendant of the Garnet	6½	10
Guy	5	9
Fall of the Garnet	4	34
Tacks	8¼	35
Sheets	6½	71
Clew Garnets	3	66
Bowlines	4¼	48
Bridles	3½	16
Pendants of Braces	4	8
Falls	3	71
Jeers	6½	96
Parrel Ropes	5	19
Leech Lines	2½	70
Futtocks	0	0
Legs of Catharpins	2½	18
Falls	2	20
Stoppers for Topsail Sheet	0	0
2 Tackles to Hoist up Shrouds	0	0
Bowsing Tackle	3	12
Tie	0	0
Halyards	0	0
MAIN TOPMAST		
Pendants of Tackle	3½	6
Falls	2¼	36
Shrouds	4½	84
Lanyards	2	28
Standing Backstays	5	120
Lanyards	2½	16
Stay	6½	24
Lanyard	3	11
Lifts	2½	66
Braces	2½	68
Pendants	3	6
Bowlines	3	50
Bridles	3	24
Pendant of the Top Rope	8	14½
Fall of the Same	5	48
Clewlines	3	87
Tie	7	9
Runner	5½	18
Halyards	3	60
Leech Lines	2½	17
Buntlines	2½	46
Futtocks	4	34
Parrel Ropes	3	10
Sheets	7½	49

	Size (ins)	Fathoms
MAIN TOPGALLANT MAST		
Stay	2	20
Braces	1½	58
Bowlines	1½	56
Bridles	1½	8
Parrel Ropes	2	3
Tie	2½	4
Halyards	2	36
Clewlines	1½	48
Lifts	1½	24
Shrouds	2	16
Lanyards	1½	8
MIZZEN MAST		
Burtons	3	6
Falls of Burtons	2½	32
Shrouds	4½	94
Lanyards	2½	26

	Size (ins)	Fathoms
Stay	6	14
Lanyard	2½	4
Halyards	0	0
Parrel Ropes	3½	7
Truss	2½	28
Sheet	3½	20
Tack	0	0
Bowlines	3	11
Brails	2½	100
Jeers	4½	36
CROSSJACK		
Lifts	0	0
Braces	2	42
Standing Lifts	3	10
Lanyards	0	0
Slings	3½	5

	Size (ins)	Fathoms
MIZZEN TOPMAST		
Shrouds	2½	30
Lanyards	1½	15
Futtocks	3	12
Braces	1½	36
Bowlines	1½	38
Bridles	1	6
Tie	2	5
Halyards	2	26
Lifts	2	30
Parrel Ropes	2	5
Stay	2½	9
Clewlines	2	42
Cat Ropes	5½	58
Pendant of the Fish Hook Rope	7	9
Fall of the Same	4	32
Stopper at the Bow	8	20

The names, sizes, and lengths for rigging a 1st Rate demonstrated.

	Size (ins)	Fathoms
Shank Painter	7	18
Stream Anchor Stopper	5	10
Shank Painter	0	0
Stoppers at the Bitts	8	9
Lanyards	$3\frac{1}{2}$	11
Viol	10	35
Pendant of the Winding Tackle	11	11
Fall of the Same	6	45
Buoy Ropes	8	60
Stream Anchor Buoy Rope	5	15
Boat Rope and Slings	8	32
Gust Rope to it	5	40
Pinnace Rope and Slings	5	40
Gust Rope to it	$4\frac{1}{2}$	40
2 Pair of Butt Slings	5	9
2 Pair of Hogshead Slings	4	8
Ordnance Slings	8	7
For Robins, Earrings, and Clew siezings for One Complete Suit of Sails	$1\frac{1}{2}$	1200
Rattling for Shrouds	$1\frac{3}{4}$	1000
Yard Lines	0	84
For Strapping of Blocks	7	50
	6	35
	5	30
	$4\frac{1}{2}$	36
	4	20

p15

	Size (ins)	Fathoms
BOWSPRIT		
1. Sheets	$4\frac{1}{2}$	61
2. Horse	$4\frac{1}{2}$	9
3. Clewlines	3	48
4. Slings for the Yard	6	6
5. Halyards	$4\frac{1}{2}$	38
6. Lifts	4	58
7. Standing Lifts	0	0
8. Lanyards	0	0
9. Pendants of Braces	4	5
10. Falls	$2\frac{1}{2}$	58
11. Wolding	$4\frac{3}{4}$	80
11. Garnets	3	58
12. Buntlines	3	61
SPRITSAIL TOPMAST		
13. Shrouds	3	26
14. Lanyards	2	13
15. Pendants of Braces	2	3
16. Falls	$1\frac{1}{2}$	42
17. Tie	$2\frac{1}{2}$	4
18. Halyards	$1\frac{1}{3}$	13
19. Clewlines	2	34
20. Pendants of Backstays	$2\frac{1}{2}$	6
21. Falls	$1\frac{1}{2}$	26
22. Lifts	1	28
23. Parrel Ropes	$1\frac{3}{4}$	4
FOREMAST		
24. Pendants	8	8
25. Falls of Tackles	$3\frac{1}{2}$	160
26. Runners of Tackles	7	30
27. Shrouds	$7\frac{3}{4}$	210
28. Lanyards	$3\frac{3}{4}$	64
29. Stay	$14\frac{1}{2}$	18
30. Collar	10	5
31. Lanyard	$4\frac{1}{2}$	14
32. Futtocks	0	0
33. Parrel Ropes	5	16
34. Clew Garnets	3	68
35. Bowlines	$4\frac{1}{2}$	64
36. Bridles	$3\frac{3}{4}$	14

	Size (ins)	Fathoms
37. Pendants of Braces	$3\frac{1}{2}$	7
38. Falls	3	58
39. Buntlines	2	69
40. Sheets	6	58
41. Tacks	$7\frac{3}{4}$	44
42. Lifts	$3\frac{1}{2}$	66
43. Leech Lines	0	0
44. Jeers	6	89
45. Legs for Catharpins	0	0
46. Falls	2	18
47. Stoppers for Topsail Sheets	0	0
48. Tie	0	0
49. Halyards	0	0
FORE TOPMAST		
50. Pendant of the Top Rope	$7\frac{1}{2}$	$14\frac{1}{2}$
51. Fall of the Same	$2\frac{1}{4}$	48
52. Shrouds	4	90
53. Lanyards	2	28
54. Standing Backstays	4	126
55. Lanyards	$2\frac{1}{4}$	18
56. Stay	$5\frac{2}{3}$	19
57. Lanyard	$3\frac{1}{2}$	12
58. Lifts	$2\frac{1}{4}$	68
59. Futtocks	3	50
60. Tie	$5\frac{3}{4}$	$9\frac{1}{2}$
61. Runner	$5\frac{1}{4}$	19
62. Halyards	$3\frac{1}{4}$	62
63. Bowlines	$2\frac{1}{2}$	80
64. Bridles	$2\frac{1}{2}$	13
65. Clewlines	3	78
66. Pendants of Tackles	$3\frac{3}{4}$	6
67. Falls	$2\frac{3}{4}$	38
68. Sheets	6	46
69. Parrel Ropes	3	10
70. Leech Lines	2	14
71. Pendants	3	5
72. Braces	$2\frac{1}{2}$	58
73. Buntlines	3	35
FORE TOPGALLANT MAST		
74. Stay	$2\frac{1}{2}$	28
75. Tie	$2\frac{1}{2}$	4
76. Halyards	$1\frac{1}{4}$	38
77. Lifts	$1\frac{1}{4}$	28
78. Braces	2	50
79. Clewlines	$1\frac{1}{2}$	60
80. Bowlines	$1\frac{1}{2}$	62
81. Bridles	$1\frac{1}{2}$	4
82. Parrel Ropes	2	4
83. Shrouds	2	20
84. Lanyards	$1\frac{1}{4}$	14
MAINMAST		
85. Pendants of Tackles	9	9
86. Runners of Tackles	$6\frac{1}{2}$	36

DEANE'S DOCTRINE OF NAVAL ARCHITECTURE, 1670

	Size (ins)	Fathoms
87. Falls of Tackles	4	124
88. Lifts	4	74
89. Shrouds	$8\frac{1}{2}$	298
90. Lanyards	4	98
91. Stay	$17\frac{1}{4}$	20
92. Collar	15	$10\frac{1}{2}$
93. Lanyard	$5\frac{1}{4}$	15
94. Pendant of the Garnet	$7\frac{1}{2}$	11
95. Guy	5	10
96. Fall of the Garnet	$4\frac{1}{4}$	43
97. Tacks	$8\frac{1}{2}$	36

	Size (ins)	Fathoms
98. Sheets	$6\frac{1}{4}$	80
99. Clew Garnets	$3\frac{1}{4}$	66
100. Bowlines	$4\frac{3}{4}$	58
101. Bridles	4	22
102. Pendants of Braces	4	10
103. Falls	3	82
104. Jeers	$7\frac{3}{4}$	115
105. Parrel Ropes	6	30
106. Leech Lines	0	0
107. Buntlines	$2\frac{3}{4}$	86
108. Futtocks	0	0

	Size (ins)	Fathoms
109. Legs of Catharpins	$2\frac{1}{2}$	18
110. Falls	2	20
111. Stoppers for Topsail Sheets	0	0
112. 2 Tackles to Hoist up Shrouds	0	0
113. Bowsing Tackle	0	0
114. Knave Line	2	20
115. Horse for the Yard	4	13

Note: The draughts show the upper yards in two positions, raised for carrying a sail, and lowered when the sail is furled. The draught of the 1st Rate has triple wales, which are highly unusual.

102

	Size (ins)	Fathoms
MAIN TOPMAST		
116. Pendants of Tackles	$4\frac{1}{2}$	6
117. Falls	$2\frac{1}{4}$	38
118. Shrouds	$4\frac{3}{4}$	104
119. Lanyards	$2\frac{1}{2}$	40
120. Standing Backstay	$4\frac{1}{2}$	156
121. Lanyards	$2\frac{1}{2}$	22
122. Stay	$6\frac{1}{2}$	26
123. Lanyard	4	12
124. Lifts	$2\frac{3}{4}$	69
125. Braces	$2\frac{1}{4}$	78
126. Pendants	$2\frac{3}{4}$	6
127. Bowlines	4	58
128. Bridles	$3\frac{1}{4}$	26
129. Pendant of the Top Rope	8	21
130. Fall of the Same	$5\frac{1}{2}$	61
131. Clewlines	$3\frac{1}{2}$	92
132. Tie	8	10
133. Runner	$5\frac{3}{4}$	22
134. Halyards	$3\frac{1}{4}$	60
135. Leech Lines	0	0
136. Buntlines	$3\frac{1}{2}$	54
137. Futtocks	5	50
138. Parrel Ropes	$3\frac{1}{4}$	12
139. Sheets	$7\frac{1}{4}$	56
MAIN TOPGALLANT MAST		
140. Stay	$2\frac{1}{2}$	21
141. Braces	$1\frac{1}{2}$	60
142. Bowlines	2	74
143. Bridles	$1\frac{1}{2}$	11
144. Parrel Ropes	2	4
145. Tie	$2\frac{1}{2}$	6
146. Halyards	2	58
147. Clewlines	$1\frac{3}{4}$	49
148. Lift	2	29
149. Shrouds	$2\frac{1}{2}$	20
150. Lanyards	$1\frac{1}{2}$	18
MIZZEN MAST		
151. Burtons	3	6
152. Falls of Burtons	2	32
153. Shrouds	$4\frac{3}{4}$	128
154. Lanyards	$2\frac{1}{4}$	46
155. Stay	$6\frac{3}{4}$	15
156. Lanyard	$2\frac{1}{2}$	6
157. Halyards	0	0
158. Parrel Ropes	$4\frac{1}{2}$	7
159. Truss	$2\frac{1}{2}$	36
160. Sheet	$3\frac{1}{4}$	21
161. Tack	3	12
162. Bowlines	$3\frac{1}{4}$	16
163. Brails	$2\frac{1}{4}$	180
164. Jeers	$4\frac{3}{4}$	40

	Size (ins)	Fathoms
CROSSJACK		
165. Lifts	0	0
166. Braces	$2\frac{1}{4}$	50
167. Standing Lifts	3	12
168. Lanyards	0	0
169. Slings	$3\frac{1}{2}$	6
MIZZEN TOPMAST		
170. Shrouds	3	36
171. Lanyards	$1\frac{1}{4}$	20
172. Futtocks	$2\frac{1}{2}$	18
173. Braces	$1\frac{3}{4}$	40
174. Bowlines	$1\frac{3}{4}$	44
175. Bridles	$1\frac{1}{2}$	9
176. Tie	3	6
177. Halyards	$2\frac{1}{2}$	32
178. Lifts	2	35
179. Parrel Ropes	$2\frac{1}{2}$	6
180. Stay	3	17
181. Clewlines	2	54
Cat Ropes	0	0
Pendants of the Fish Hook Rope	$7\frac{1}{2}$	13
Fall of the Same	4	32
Stoppers at the Bow	$6\frac{1}{2}$	22
Shank Painter	$6\frac{1}{2}$	24
Stream Anchor Stopper	5	11
Shank Painter	5	10
Stoppers at the Bitts	9	16
Lanyards	$3\frac{1}{2}$	12
Viol	9	35
Pendant of the Winding Tackle	11	12
Fall of the Same	12	11
Buoy Ropes	$7\frac{1}{2}$	90
Stream Anchor Buoy Rope	5	16
Boat Rope and Slings	8	40
Gust Rope to it	5	46
Pinnace Rope and Slings	$5\frac{1}{2}$	46
Gust Rope to it	4	40
2 Pair of Butt Slings	5	10
2 Pair of Hogshead Slings	4	8
Ordnance Slings	8	7
For Robins, Earrings and Clew Siezings for one Complete Suit of Sails	$1\frac{1}{2}$	1250
Rattling for Shrouds	$1\frac{3}{4}$	1100
Yard Lines	0	90
Strapping for Blocks	$7\frac{3}{4}$	56
	7	40
	6	34
	$4\frac{3}{4}$	38
	4	22

The price of the ship's hull, the number of all stores and men, rigging, victualling, and wages, and full charge for six months ready fixed for sea.

p85 The reason of the different prices which you find in the margin in the table of lost ships proceeds from the materials on board at the present time they were lost, and also the nature of their guns, some ships having all brass guns, other but some brass guns, some without any guns at all. This, being considered, is very nigh calculated in every respect, which I thought good to signify for your more clear inspection. There is other great charges in this late Dutch War, as transportation of all provision, loss of fireships, sunk ships in harbours, and the like, all which, being without my reach, am constrained to leave it to those who can come nearer the truth.

	£	s	d
Royal Charles	22971	10	0
Royal James	19860	0	0
Royal Oak	19172	0	0
London	50851	0	0
St Andrew	58877	0	0
Prince	63506	0	0
Swiftsure	32263	0	0
Vanguard	8965	0	0
Essex	12860	0	0
Resolution	25600	0	0
Defiance	6578	0	0
Convertine	10053	0	0
Elizabeth	7905	1	0
Mathias	9625	0	0
Nonsuch	6075	4	0
Breda	8045	0	0
Phoenix	6272	0	0
Sapphire	6425	0	0
St Patrick	8230	0	0
Colchester	4788	6	0
Convert	8862	0	0
Oxford	4626	2	0
Martin	1626	0	0

pp77-78

	Number of Men	Number of Guns	Length by the Keel (ft ins)	Breadth by the Beam (ft ins)	Depth in Hold (ft ins)	Draught of Water (ft ins)	Number of Tons	Tuns and Tunnage	When Built	By Whom Built	Where Built	Weight of Cordage to Rig (tons cwt lbs)	Number of Cables	Weight of Cables (tons cwt)	Number of Anchors	Weight of Anchors (tons cwt)	Number of Blocks and Deadeyes	Number of Sails	Yards of Canvas	Number of Boats	Barrels of Powder	Weight of Shot (tons cwt)	Weight of Guns (tons cwt)	Price of the Hull complete from the Builder (£ s d) [*cost per ton]
Royal Sovereign	800	100	127-0	47-0	19-0	22-4	1492	1989	1637	Phi Pett	Woolwich	28-0-0	14	45-4	10	18-18	554	34	11617	3	400	54-0	176-0	29840-0 *£20
Charles	600	96	128-0	42-6	18-6	21-0	1229	1638	1668	Mr Shish	Deptford	19-2-0	14	35-16	9	14-12	496	24	8984	3	370	50-0	160-0	18435-0 *£15
Royal James	500	84	122-0	42-0	18-0	21-0	1121	1480	1658	Chr Pett	Woolwich													
Royal James	800	104	132-6	45-0	18-0	19-9	1416	1878	1670	Capt Deane	Portsmouth	20-12-2	14	39-1	9	15-10	530	24	9764	3	400	56-0	178-0	24072-0 *£17
Prince	800	102	131-0	44-10	19-0	20-9	1403	1870	1670	Phi Pett	Chatham	20-10-0	14	39-6	9	15-15	530	24	9764	3	400	54-0	177-13	23851-0 *£17
St Andrew	600	96	129-0	43-6	18-8	21-0	1298	1730	1670	Chr Pett	Woolwich	19-2-0	14	36-12	9	14-18	530	24	9010	3	380	51-0	161-6	21417-0 *£16-10

Note: These tables give details of all the ships in the Royal Navy in 1670. They agree substantially with the lists made by Pepys, though the latter gives only dates, dimensions, and builders. (Navy Records Society, Vol XXVI, pages 256-311). A list very similar to Deane's is to be found in a manuscript in the National Maritime Museum (RUSI/NM/106), which is conventionally dated from the time of James II (1685-8), but in fact belongs to this period.

The ships are divided into Rates, though this is not made explicit in the document. The divisions are as follows:

First Rate. *Royal Sovereign* to *St Andrew*.
Second Rate. *Loyal London* to *St Michael*.
Third Rate. *Resolution* to *Montague*.
Fourth Rate. *Greenwich* to *Nonsuch*.
Fifth Rate. *Falcon* to *Little Victory*.
The last group, from *Drake* to *Saudadoes,* is a miscellaneous collection of 6th Rates, sloops, and yachts.

The list brings out the lack of uniformity in the fleet. It is difficult to find two ships with the same rate which would require the same complement of rigging, stores, guns, etc, and this must have augmented the supply problem. The only feature common within each Rate is the 'Charge of Officers, 6 Months', and this emphasises that the main and original reason for dividing the ships into Rates was to provide an establishment for officers' numbers and rates of pay.

Several ships are listed, but have no details given after their names. This is because they had been lost in recent years. For example, the *Loyal London* was burnt by the Dutch in 1667, and the *Nonsuch* cast away in 1664.

The following price rates are also given in the tables:
Cordage   at £45 per ton.
Cables   at £45 per ton.
Sails   17d per yard.
Brass Guns £150 per ton.
Iron Guns   £18 per ton.
Shot   '11 and 40 per ton'.
Powder   £3 per ton.
Victualling 8d per day.
Seamen   23s. per month.

It is noted in the margin: 'Memorandum — Anchors are of several prices according to their bigness, the medium price near as in the margin.' Thus prices of anchors vary according to the size of the ship, as follows:

Sovereign and *Royal Charles* £3.17s per ton
Other First Rates and
Second Rates   £3.5s. per ton
Third Rates   £2.6s. per ton
Fourth Rates   £2.3s. per ton
Fifth Rates and below   £2 per ton

Other contemporary sources give a rather different account of the sails carried by each ship. A list dated 1670 in the National Maritime Museum [RCE/3] gives the following sails for a typical ship:

Spritsail Courses	3
Spritsail Topsails	2
Fore Courses	3
Fore Bonnets	1
Fore Topsails	3
Fore Topgallants	2
Main Courses	3
Main Bonnets	1
Main Topsails	3
Main Topgallants	2
Mizzen Courses	3
Mizzen Topsails	2
Main Staysails	2
Main Top Staysails	2
Fore Top Staysails	2
Main Studding Sails	2
Main Top Studding Sails	2

A total of 40 sails, 20 of which were spare.

An undated list, which may refer to the thirty ships of 1677, [British Library, cuf 651e (28-31)] gives the following:

Spritsail Courses	2
Spritsail Topsail	1
Fore Courses	2
Fore Bonnet	1
Fore Topsails	2
Fore Topgallant Sail	1
Main Courses	2
Main Bonnet	1
Main Topsails	2
Main Topgallant Sail	1
Mizzen Courses	2
Mizzen Bonnet	1
Mizzen Topsails	2
Main Staysail	1
Mizzen Staysail	1
Main Topmast Staysail	1
Fore Topmast Staysail	1
Main Studding Sails	2
Topsail Studding Sails	2

A total of 28 sails of which 7 were spare. Deane makes no mention of either studding sails or staysails in his rigging lists.

Prices in the tables are given in pounds and shillings, and sometimes pence, and weights in tons, hundredweight, and pounds.

The building place of the *Merlin* is given as Rotherhithe in Pepys' and other contemporary lists and it seems that the area of Rotherhithe around the present Southwark Park was then known as 'Jamaica Level'.

Price of Cordage to Rig (£ s)	Price of Cables (£ s)	Price of Anchors (£ s d)	Price of Blocks, Tops, and Pumps (£ s)	Price of Sails (£ s d)	Price of Brass Guns (£ s)	Price of Iron Guns (£ s)	Price of Shot (£ s)	Price of Powder (£ s)	Price of Gunner's Stores New (£ s)	Price of Boatswain's Stores and Cordage (£ s)	Price of Carpenter's Stores (£ s)	Price of the Ship Completely Rigged and Stored (£ s d)	Charge of Victualling, 6 Months (£ s d)	Charge of Seamen, 6 Months (£ s d)	Charge of Officers, 6 Months (£ s d)	Charge of the Whole Ship Completely Rigged, Stored and Victualled and Wages for 6 Months (£ s d)
1260-0	2034-0	1423-0	230-0	831-13-4	26400-0	–	728-0	1200-0	1678-0	800-0	92-0	66506-13-4	4853-6-8	5598-15-6	869-0	77827-15-6
859-10	1611-0	975-4	215-0	690-19-0	24000-0	–	650-0	1110-0	1578-0	725-0	92-0	50851-13-0	3640-0-0	4103-15-6	869-0	59464-8-6
928-2	1757-5	1084-0	220-0	802-0-0	26550-0	–	750-0	1200-0	1587-0	725-0	92-0	59767-7-0	4853-6-8	5598-15-6	869-0	71088-9-2
922-10	1768-10	1095-0	220-0	802-0-0	26500-0	–	720-0	1200-0	1587-0	725-0	92-0	59483-0-0	4853-6-8	5598-15-6	869-0	70804-2-2
859-10	1647-0	1050-0	220-0	750-0-0	24200-0	–	695-0	1140-0	1578-0	725-0	92-0	54373-10-0	3640-0-0	4103-15-6	869-0	62986-5-6

DEANE'S DOCTRINE OF NAVAL ARCHITECTURE, 1670

	Number of Men	Number of Guns	Length by the Keel (ft ins)	Breadth by the Beam (ft ins)	Depth in Hold (ft ins)	Draught of Water (ft ins)	Number of Tons	Tuns and Tunnage	When Built	By Whom Built	Where Built	Weight of Cordage to Rig (tons cwt lbs)	Number of Cables	Weight of Cables (tons cwt)	Number of Anchors	Weight of Anchors (tons cwt)	Number of Blocks and Deadeyes	Number of Sails	Yards of Canvas	Number of Boats	Barrels of Powder	Weight of Shot (tons cwt)	Weight of Guns (tons cwt)	Price of the Hull complete from the Builder (£ s d) (*cost per ton)
Loyal London	600	96	129-0	43-9	19-0	21-0	1312	1749	1670	Mr Shish	Deptford	19-2-0	14	36-10	9	14-13	496	24	8984	3	370	50-0	160-0	19680-0-0 *£15
Loyal London										Capt Taylor	Chatham													
Royal Katherine	480	82	120-0	40-0	17-4	20-8	1021	1361	1664	Chr Pett	Woolwich	15-18-3	12	29-16	9	12-4	486	24	7244	3	320	45-0	140-0	14804-10- *£14-10
Henry	400	74	120-0	38-4	15-9	20-8	920	1220	1656	Mr Callis	Deptford	15-0-0	12	29-16	9	12-4	486	24	6984	3	310	41-0	135-0	13340-0-0
Triumph	380	70	117-0	38-6	15-6	18-0	922	1229	1623	Mr Burrell	Deptford	14-14-0	11	24-18	8	9-4	486	24	6984	3	275	37-0	110-0	10142-0-0 *£11
Rainbow	350	66	114-0	36-6	15-0	17-6	807	1076	1617	Mr Bright	Deptford	13-16-0	9	20-11	8	9-4	486	24	6984	2	270	37-0	108-0	8877-0-0
James	360	68	116-0	39-0	16-0	18-6	938	1250	1633	Capt Pett	Deptford	14-14-0	9	24-17	8	9-4	486	24	6984	2	275	37-0	110-0	11256-0-0 *£12
George	330	64	117-0	38-9	15-9	18-6	937	1249	1622	Mr Burrell	Deptford	14-14-0	9	24-16	8	8-4	486	24	6984	2	270	35-0	104-0	10307-0-0 *£11
Victory	480	84	124-0	42-0	17-6	20-6	1168	1551	1666	Phi Pett	Chatham	17-0-0	13	34-10	8	12-18	486	24	7256	3	360	45-0	145-0	16870-15 *£14-10
Unicorn	330	64	110-0	35-8	16-0	17-6	730	973	1633	Mr Boate	Woolwich	14-0-0	9	20-11	8	8-1	486	24	6984	2	270	35-0	104-0	8030-0-0 *£11
St Michael	480	84	122-6	40-0	17-5	20-0	1042	1389	1669	Mr Tippets	Portsmouth	15-18-3	12	29-16	9	12-4	486	24	7244	3	320	44-0	140-0	15630-0-0 *£15
Resolution	320	70	120-6	37-2	15-9	16-9	885	1180	1667	Capt Deane	Harwich	10-18-2	8	17-7	6	7-12	484	24	6892	2	300	37-0	118-0	7965-0- *£9
Warspite	320	68	117-0	38-0	15-4	17-6	898	1197	1666	Mr Johnson	Blackwall	10-18-2	8	17-5	6	7-12	484	24	6892	2	300	37-0	110-0	6510-10 *£7-5
Cambridge	320	68	121-0	37-0	15-9	17-6	881	1174	1666	Mr Shish	Deptford	10-18-2	8	17-8	6	7-12	484	24	6892	2	300	37-0	110-0	7929-0- *£9
Defiance	320	66	116-0	37-0	15-6	17-6	823	1097	1665	Mr Castle	Deptford	10-18-2	8	16-14	6	7-12	484	24	6892	2	300	37-0	110-0	5966-15 *£7-5
Rupert	320	68	119-0	36-3	15-6	17-1	791	1054	1665	Capt Deane	Harwich	10-18-2	8	16-14	6	7-12	484	24	6892	2	300	37-0	110-0	7119-0- *£9
Edgar	320	70	124-0	39-10	16-0	18-4	1055	1406	1668	Mr Bayley	Bristol	10-18-2	8	17-8	6	7-19	484	24	6892	2	300	37-0	112-0	7680-15 *£7-5
Monmouth	320	66	119-0	36-0	15-6	18-0	822	1082	1655	Chr Pett	Chatham	10-18-2	8	16-14	6	7-12	484	24	6892	2	290	37-0	110-0	7398-0- *£9
Fairfax	320	66	120-0	35-2	14-6	17-6	785	1046	1649	Capt Taylor	Chatham	10-18-2	8	16-14	6	7-12	484	24	6892	2	275	37-0	104-0	7065-0-
Dunkirk	250	56	112-0	32-6	14-0	16-6	629	838	1651	Mr Burrell	Woolwich	10-0-0	8	14-13	6	6-12	484	24	6269	2	270	30-0	96-0	5346-10 *£8-10
Dreadnought	250	58	116-8	34-6	14-2	17-0	738	984	1654	Mr Johnson	Blackwall	10-10-0	8	15-7	6	6-10	484	24	6618	2	270	35-0	104-0	5350-10 *£7-5
Mary	280	60	116-0	34-8	14-6	17-6	741	988	1640	Chr Pett	Woolwich	10-10-0	8	16-2	6	6-10	484	24	6718	2	275	35-0	104-0	6669-0- *£9
Plymouth	260	60	116-0	34-8	14-6	17-6	741	988	1654	Capt Taylor	Wapping	10-10-0	8	16-2	6	6-12	484	24	6644	2	275	35-0	104-0	5372-5- *£7-5
Revenge	260	60	117-6	35-0	14-5	17-6	746	984	1654	Mr Graves	Limehouse	10-10-0	8	16-2	6	6-12	484	24	6558	2	275	35-0	104-0	5408-10

RIGGING

Price of Cordage to Rig (£ s)	Price of Cables (£ s)	Price of Anchors (£ s d)	Price of Blocks, Tops, and Pumps (£ s)	Price of Sails (£ s d)	Price of Brass Guns (£ s)	Price of Iron Guns (£ s)	Price of Shot (£ s)	Price of Powder (£ s)	Price of Gunner's Stores New (£ s)	Price of Boatswain's Stores and Cordage (£ s)	Price of Carpenter's Stores (£ s)	Price of the Ship Completely Rigged and Stored (£ s d)	Charge of Victualling, 6 Months (£ s d)	Charge of Seamen, 6 Months (£ s d)	Charge of Officers, 6 Months (£ s d)	Charge of the Whole Ship Completely Rigged, Stored and Victualled and Wages for 6 Months (£ s d)
859-10	1611-0	975-4	215-0	639-0-0	24000-0	–	650-0	1110-0	1578-0	700-0	92-0	53071-13-0	3640-0-0	4178-10-6	685-12	61755-15-6
717-3	1341-0	753-16	180-0	513-2-4	21000-0	–	630-0	960-0	1037-7	461-0	61-0	42458-8-4	2912-0-0	3281-10-6	685-12	49338-0-10
675-0	1341-0	753-16	180-0	494-14-0	20250-0	–	580-0	930-0	1037-7	461-0	61-0	40103-17-0	2426-13-4	2683-10-6	685-12	46982-19-6
661-10	1120-10	535-16	180-0	494-14-0	16500-0	–	507-0	825-0	789-10	461-0	61-0	32278-0-0	2305-6-8	2534-0-6	685-12	37802-19-2
621-0	924-15	535-16	170-0	494-14-0	16200-0	–	507-0	810-0	789-10	400-0	61-0	30290-15-0	2123-6-8	2309-15-6	685-12	35409-9-2
661-10	1118-5	535-16	180-0	494-14-0	16500-0	–	507-0	825-0	789-10	461-0	61-0	33389-15-0	2184-0-0	2384-10-6	685-12	38643-17-6
661-11	1116-0	453-17	170-0	494-14-0	15600-0	–	400-0	810-0	789-10	400-0	61-0	31263-11-0	2002-0-0	2160-5-6	685-12	36111-8-6
765-0	1552-10	783-0	180-0	513-19-4	21750-0	–	630-0	1080-0	1489-0	500-0	80-0	46194-4-4	2912-0-0	3281-10-6	685-12	53073-6-10
630-0	924-15	445-17	170-0	494-14-0	15600-0	–	400-0	810-0	798-0	400-0	61-0	28155-6-0	2002-0-0	2160-5-6	685-12	33598-3-6
717-3	1341-0	753-16	180-0	513-2-0	21000-0	–	619-0	960-0	1037-7	461-0	61-0	43273-8-0	2912-0-0	3281-10-6	685-12	50151-8-6
491-12	780-15	350-4	128-0	488-3-8	–	2124-0	507-0	900-0	789-10	385-0	48-0	14957-4-8	1941-6-8	2152-16-0	646-7	19697-14-4
491-12	776-5	350-4	128-0	488-3-8	–	1980-0	507-0	900-0	789-10	385-0	48-0	13354-4-8	1941-6-8	2152-16-0	646-7	18094-14-4
491-12	783-0	350-4	128-0	488-3-8	–	1980-0	507-0	900-0	789-10	385-0	48-0	14799-9-8	1941-6-8	2152-16-0	646-7	19539-19-4
491-12	751-10	350-4	128-0	488-3-8	–	1980-0	507-0	900-0	789-10	385-0	48-0	12777-14-8	1941-6-8	2152-16-0	646-7	17518-4-4
491-12	751-10	350-4	128-0	488-3-8	–	1980-0	507-0	900-0	789-10	385-0	48-0	13937-19-8	1941-6-8	2152-16-0	646-7	18678-9-4
491-12	783-0	399-10	128-0	488-3-8	–	2016-0	507-0	900-0	789-10	385-0	48-0	14616-10-8	1941-6-8	2152-16-0	646-7	19356-19-4
491-12	751-10	350-4	128-0	488-3-8	–	1980-0	507-0	870-0	789-10	385-0	48-0	14186-19-8	1941-6-8	2152-16-0	646-7	18927-9-4
491-12	751-10	350-4	128-0	488-3-8	–	1872-0	507-0	825-0	789-10	385-0	48-0	13690-19-8	1941-6-8	2152-16-0	646-7	18431-9-4
450-0	659-5	300-5	128-0	441-1-1	–	1728-0	400-0	810-0	789-10	385-0	48-0	12488-11-1	1516-13-4	1629-11-0	646-7	16301-2-5
472-10	690-15	295-14	128-0	468-15-6	–	1872-0	475-0	810-0	789-10	385-0	48-0	11785-14-6	1516-13-4	1629-11-0	646-7	15578-5-10
472-10	724-10	295-14	128-0	475-17-2	–	1872-0	475-0	825-0	789-10	385-0	48-0	12860-1-2	1698-10-4	1853-16-0	646-7	17058-17-6
472-10	724-10	300-5	128-0	470-12-4	–	1872-0	475-0	825-0	789-10	385-0	48-0	11862-12-4	1577-6-8	1704-6-0	646-7	15790-12-0
472-10	724-10	300-5	128-0	464-10-0	–	1872-0	475-0	825-0	789-10	385-0	48-0	11892-15-6	1577-6-8	1704-6-0	646-7	15820-15-2

DEANE'S DOCTRINE OF NAVAL ARCHITECTURE, 1670

pp79-80

	Number of Men	Number of Guns	Length by the Keel (ft ins)	Breadth by the Beam (ft ins)	Depth in Hold (ft ins)	Draught of Water (ft ins)	Number of Tons	Tuns and Tunnage	When Built	By Whom Built	Where Built	Weight of Cordage to Rig (tons cwt lbs)	Number of Cables	Weight of Cables (tons cwt)	Number of Anchors	Weight of Anchors (tons cwt)	Number of Blocks and Deadeyes	Number of Sails	Yards of Canvas	Number of Boats	Barrels of Powder	Weight of Shot (tons cwt)	Weight of Guns (tons cwt)	Price of the Hull complete from the Builder (£ s d) (*cost per ton)
York	250	56	116-0	34-6	14-2	17-0	734	978	1654	Mr Johnson	Blackwall	10-10-0	8	16-2	6	6-8	484	24	6418	2	270	35-0	104-0	5321-10- *£7-5
Ann	250	56	116-9	34-7	14-2	17-0	742	989	1654	Mr Chamberlain	Deptford	10-10-0	8	16-2	6	6-12	482	24	6558	2	270	35-0	104-0	5379-10-
Gloucester	250	56	117-0	34-10	14-6	18-0	755	1007	1654	Mr Graves	Limehouse	10-10-0	8	16-2	6	6-13	482	24	6558	2	270	35-0	104-0	5473-15-
Henrietta	260	60	116-0	35-7	14-4	17-6	768	1024	1654	Mr Bright	Horsleydown	10-18-0	8	16-9	6	6-19	482	24	6647	2	275	35-0	104-0	5568-0-0
Lion	260	60	112-0	35-0	16-6	17-0	728	970	1640	Mr Apslin	Chatham	10-10-0	8	14-12	6	6-10	482	24	5248	2	275	35-0	104-0	6552-0-0 *£9
Monck	250	56	107-0	35-0	14-6	17-6	697	929	1659	Mr Tippets	Portsmouth	10-0-0	8	14-12	6	6-4	482	24	5248	2	270	30-0	100-0	6273-0-0
Montague	320	66	117-0	35-2	15-0	17-9	780	1040	1654	Mr Tippets	Portsmouth	10-18-0	8	16-4	6	7-10	482	24	6892	2	275	35-0	104-0	7020-0-0
Greenwich	250	60	110-0	33-6	15-0	17-0	646	861	1666	Chr Pett	Woolwich	9-0-0	8	14-11	6	6-3	482	24	5258	2	210	22-0	94-0	5491-0-0 *£8-10
Leopard	250	58	109-0	33-9	15-0	17-3	645	860	1667	Mr Shish	Deptford	9-0-0	8	14-11	6	6-3	482	24	5258	2	210	22-0	94-0	5482-10
Assurance	160	42	87-0	27-0	11-0	12-6	337	449	1646	Pet Pett	Deptford	6-10-0	7	10-8	6	4-12	475	24	3072	2	130	13-0	50-0	2352-0-0 *£7
Assistance	170	48	102-0	31-0	13-0	15-0	521	694	1650	Mr Johnson	Deptford	7-2-3	7	11-0	6	4-17	475	24	3264	2	130	14-10	54-0	3386-10 *£6-10
Adventure	150	40	94-0	27-9	13-10	13-9	374	498	1646	Pet Pett Jnr	Woolwich	6-10-0	7	10-1	6	4-10	475	24	3264	2	125	13-0	44-0	2618-0-0
Advice	180	50	100-0	31-2	15-7	16-0	513	684	1650	Pet Pett Jnr	Woolwich	7-3-2	7	11-1	6	4-18	475	24	4072	2	130	14-10	54-0	3334-10
Bristol	180	52	104-0	31-1	15-6	15-6	532	709	1653	Mr Tippets	Portsmouth	7-3-2	7	11-8	6	5-0	475	24	4072	2	135	14-10	54-0	4256-0-0 *£8
Centurion	170	48	104-0	31-0	15-0	16-6	531	708	1650	Pet Pett Jnr	Ratcliff	7-3-2	7	11-0	6	5-0	475	24	4072	2	130	14-10	54-0	3451-10 *£6-10
Diamond	170	48	105-0	31-3	15-7	16-0	545	726⅔	1651	Pet Pett Jnr	Deptford	7-3-2	7	11-0	6	5-0	475	24	3826	2	135	14-10	54-0	4360-0-0 *£8
Dover	170	48	104-0	31-8	15-10	17-0	554½	738⅔	1650	Mr Castle	Redriffe	7-3-2	7	11-0	6	5-0	475	24	3264	2	130	14-10	54-0	3604-5-0 *£6-10
Dragon	160	40	96-0	28-6	14-3	15-0	422	562	1647	Mr Goddard	Chatham	6-10-0	7	10-0	6	4-16	475	24	3826	2	130	13-0	44-0	2473-0-
Foresight	170	48	102-0	31-1	13-0	14-6	522	696	1650	Jonas Shish	Deptford	7-3-2	7	11-0	6	4-19	475	24	3826	2	130	14-10	54-0	3393-0-
Swallow	170	46	100-10	31-10	13-1	14-1	543	724	1653	Tho Taylor	Pitchouse	7-3-2	7	11-0	6	4-18	475	24	3826	2	130	14-10	54-0	3529-10
Hampshire	170	48	101-9	29-9	14-10	14-10	479	638	1653	Phi Pett	Deptford	7-3-2	7	11-0	6	4-18	475	24	3826	2	130	14-10	54-0	3592-10 *£7-10
Jersey	180	50	101-10	32-2	13-2	14-0	560	746	1654	Mr Starling	Essex	7-3-2	7	11-0	6	4-18	475	24	3826	2	135	14-10	54-0	3640-0-0 *£6-10
Kent	180	52	107-0	32-6	13-0	15-0	601	801	1652	Mr Johnson	Deptford	7-3-2	7	11-0	6	4-18	475	24	3826	2	135	14-10	56-0	3906-10
Mary Rose	170	48	100-0	31-8	13-0	15-0	528	704	1654	Mr Carey	Woodbridge	7-3-2	7	11-0	6	4-18	475	24	3826	2	135	14-10	54-0	3432-0-
Marmaduke	170	46	94-0	29-6	14-0	16-0	400	553	1643			7-3-0	7	11-0	6	4-18	475	24	3826	2	130	14-10	54-0	2600-0-
Newcastle	240	56	108-6	33-1	13-0	15-0	631	841	1653	Phi Pett	Ratcliff	8-0-0	7	11-0	6	4-18	475	24	3826	2	200	14-10	54-0	4101-10
Princess	180	50	105-0	31-6	14-6	17-0	556	741	1661	Mr Furzer	Forest	7-3-2	7	11-0	6	4-18	475	24	3826	2	135	14-10	54-0	4726-0- *£8-10
Nonsuch										Pet Pett Jnr														
Portsmouth	180	48	99-0	28-4	14-2	15-0	422	562	1649	Mr Eastwood	Portsmouth	7-3-2	7	11-0	6	4-18	475	24	3826	2	130	14-10	54-0	3587-0-
Portland	180	50	105-0	32-0	12-10	15-0	605	806	1653	Capt Taylor	Wapping	7-3-2	7-	11-0	6	4-18	475	24	3826	2	135	14-10	54-0	3932-10 *£6-10

RIGGING

Price of Cordage to Rig (£ s)	Price of Cables (£ s)	Price of Anchors (£ s d)	Price of Blocks, Tops, and Pumps (£ s)	Price of Sails (£ s d)	Price of Brass Guns (£ s)	Price of Iron Guns (£ s)	Price of Shot (£ s)	Price of Powder (£ s)	Price of Gunner's Stores New (£ s)	Price of Boatswain's Stores and Cordage (£ s)	Price of Carpenter's Stores (£ s)	Price of the Ship Completely Rigged and Stored (£ s d)	Charge of Victualling, 6 Months (£ s d)	Charge of Seamen, 6 Months (£ s d)	Charge of Officers, 6 Months (£ s d)	Charge of the Whole Ship Completely Rigged, Stored and Victualled and Wages for 6 Months (£ s d)
472-10	724-10	290-14	128-0	454-12-2	–	1872-0	475-0	810-0	789-10	385-0	48-0	1171-6-2	1516-13-4	1629-11-0	646-7	15563-17-6
472-10	724-10	305-0	128-0	464-10-6	–	1872-0	475-0	810-0	789-10	385-0	48-0	11848-15-6	1516-13-4	1629-11-0	646-7	15641-6-10
472-10	724-10	307-15	128-0	464-10-6	–	1872-0	475-0	810-0	789-10	385-0	48-0	11950-10-6	1516-13-4	1629-11-0	646-7	15743-1-10
490-10	740-5	321-0	128-0	470-16-7	–	1872-0	475-0	825-0	789-10	385-0	48-0	12113-1-7	1577-6-8	1704-6-0	646-7	16041-1-3
472-10	657-0	291-0	128-0	371-14-8	–	1872-0	475-0	825-0	789-10	385-0	48-0	12868-14-8	1577-6-8	1704-6-0	646-7	16796-14-4
450-0	657-0	260-0	128-0	371-14-8	–	1800-0	400-0	810-0	789-10	385-0	48-0	12372-4-8	1516-13-4	1629-11-0	646-7	16164-16-0
490-10	751-10	349-10	128-0	488-3-2	–	1872-0	475-0	825-0	789-10	385-0	48-0	13616-3-2	1941-6-8	2152-16-0	646-7	18356-12-10
405-0	654-15	251-0	85-0	372-8-10	–	1692-0	329-0	630-0	578-19	300-0	40-0	10829-2-10	1516-13-4	1644-10-0	432-15	14423-1-2
405-0	654-15	251-0	85-0	372-8-10	–	1692-0	329-0	630-0	578-19	300-0	40-0	10820-12-10	1516-13-4	1644-10-0	432-15	14414-11-2
292-10	468-0	180-14	79-0	217-12-0	–	900-0	176-0	390-0	514-7	244-0	32-0	5846-3-0	970-13-4	771-15-0	432-15	8221-6-4
321-3	495-0	193-14	79-0	231-4-0	–	972-0	192-0	390-0	514-7	244-0	32-0	7050-18-0	1031-6-8	1046-10-0	432-15	9561-9-8
292-10	452-5	175-17	79-0	212-10-0	–	792-0	176-0	375-0	488-10	244-0	32-0	5937-12-0	910-0-0	897-0-0	432-15	8877-7-0
322-17	497-5	195-4	79-0	231-4-0	–	972-0	192-0	390-0	514-7	244-0	32-0	7005-7-0	1092-0-0	1121-5-0	432-15	9651-7-0
322-17	513-0	210-0	79-0	288-8-8	–	972-0	192-0	405-0	514-7	244-0	32-0	8019-2-8	1092-0-0	1121-5-0	432-15	10665-2-8
322-17	495-0	210-0	79-0	288-8-8	–	972-0	192-0	390-0	514-7	244-0	32-0	7177-12-8	1031-6-8	1046-10-0	432-15	9688-4-4
322-17	495-0	210-0	79-0	288-8-8	–	972-0	192-0	405-0	514-7	244-0	32-0	8095-2-8	1031-6-8	1046-10-0	432-15	10605-14-4
322-17	495-0	210-0	79-0	271-0-2	–	972-0	192-0	390-0	514-7	244-0	32-0	7316-19-2	1031-6-8	1046-10-0	432-15	9827-10-0
292-10	450-0	180-0	79-0	231-4-0	–	792-0	176-0	390-0	488-10	244-0	32-0	6091-4-0	970-13-4	971-15-0	432-15	8473-7-4
322-17	495-0	198-8	79-0	271-0-2	–	972-0	192-0	390-0	514-7	244-0	32-0	7102-12-2	1031-6-8	1046-10-0	432-15	9613-3-10
322-17	495-0	196-4	79-0	271-0-2	–	972-0	192-0	390-0	514-7	244-0	32-0	7350-18-2	1031-6-8	1046-10-0	432-15	9861-9-10
322-17	495-0	196-4	79-0	271-0-2	–	972-0	192-0	390-0	514-7	244-0	32-0	7300-18-2	1031-6-8	1046-10-0	432-15	9811-9-10
322-17	495-0	196-4	79-0	271-0-2	–	972-0	192-0	405-0	514-7	244-0	32-0	7363-8-2	1092-0-0	1121-5-0	432-15	10009-9-2
322-17	495-0	196-4	79-0	271-0-0	–	1008-0	192-0	405-0	514-7	244-0	32-0	7665-18-2	1092-0-0	1121-5-0	432-15	10311-18-2
322-17	495-0	196-4	79-0	271-0-0	–	972-0	192-0	405-0	514-7	244-0	32-0	7157-8-2	1031-6-8	1046-10-0	432-15	9767-19-10
321-15	495-0	196-4	79-0	271-0-0	–	972-0	192-0	390-0	514-7	244-0	32-0	6307-6-2	1031-6-8	1046-10-0	432-15	8917-17-10
360-0	495-0	196-4	79-0	271-0-2	–	972-0	192-0	600-0	514-7	244-0	32-0	8051-1-0	1456-0-0	1569-15-0	432-15	11515-15-0
322-17	495-0	196-4	79-0	271-0-2	–	972-0	192-0	405-0	514-7	244-0	32-0	8449-8-2	1092-0-0	1121-5-0	432-15	11095-8-2
322-17	495-0	196-4	79-0	271-0-2	–	972-0	192-0	390-0	514-7	244-0	32-0	7295-8-2	1092-0-0	1121-5-0	432-15	9941-8-2
322-17	495-0	196-4	79-0	271-0-2	–	972-0	192-0	405-0	514-7	244-0	32-0	7555-18-2	1092-0-0	1121-5-0	432-15	10401-18-2

DEANE'S DOCTRINE OF NAVAL ARCHITECTURE, 1670

pp81-82

	Number of Men	Number of Guns	Length by the Keel (ft ins)	Breadth by the Beam (ft ins)	Depth in Hold (ft ins)	Draught of Water (ft ins)	Number of Tons	Tuns and Tunnage	When Built	By Whom Built	Where Built	Weight of Cordage to Rig (tons cwt lbs)	Number of Cables	Weight of Cables (tons cwt)	Number of Anchors	Weight of Anchors (tons cwt)	Number of Blocks and Deadeyes	Number of Sails	Yards of Canvas	Number of Boats	Barrels of Powder	Weight of Shot (tons cwt)	Weight of Guns (tons cwt)	Price of the Hull complete from the Builder (£ s d) (*cost per ton)
Antelope	180	50	101-0	30-0	14-1	16-0	483½	644	1654	Jon Munday	Woodbridge	7-3-2	7	11-0	6	4-18	475	24	3826	2	135	14-10	54-0	3142-15
Bonaventure	180	48	100-0	29-0	12-6	15-0	450½	600	1649	Pet Pett	Deptford	7-3-2	7	11-0	6	4-18	475	24	3826	2	131	14-10	54-0	3378-15 *£7-8
Ruby	180	48	105-6	31-6	15-9	16-0	556¾	741½	1651	Pet Pett Snr	Deptford	7-3-2	7	11-1	6	5-0	438	24	3826	2	135	14-10	54-0	4175-12 *£17-10
Reserve	180	48	100-0	31-1	15-6	16-0	516	684	1650	Pet Pett Jnr	Woodbridge	7-3-2	7	11-1	6	4-18	438	24	3826	2	135	14-10	54-0	3334-10 *£6-10
Sapphire	160	46	100-0	28-10	11-9	13-0	442	589	1651	Pet Pett Sen	Ratcliffe	7-0-0	7	11-0	6	4-18	438	24	3826	2	130	14-10	50-0	2873-0-
Crown	180	48	106-0	31-8	13-0	14-6	536	714	1654	Mr Castle	Rederif	7-3-2	7	11-1	6	4-18	438	24	4000	2	135	14-10	54-0	3484-0-
Tiger	160	44	99-0	29-4	14-0	14-9	448	597	1647	Pet Pett Sen	Deptford	6-18-0	7	11-0	6	4-18	438	24	3826	2	130	13-0	54-0	2912-0-
Happy Return	180	50	104-0	33-2	13-0	18-0	605	806	1654	Mr Edgar	Yarmouth	7-3-2	7	11-0	6	4-18	438	24	4072	2	135	14-10	54-0	3932-1
Warwick	170	46	85-0	26-0	10-6	12-0	305	406	1648	Pet Pett	Ratcliffe	7-0-0	7	10-7	6	4-12	438	24	3600	2	125	14-10	44-0	1982-10
Yarmouth	180	50	105-0	33-0	13-3	17-9	608	810	1653	Mr Edgar	Yarmouth	7-3-2	7	11-0	6	4-18	438	24	3800	2	135	14-10	54-0	3952-0-
Nonsuch	160	44	88-3	27-8	10-10	12-8	359	478	1668	Capt Deane	Portsmouth	6 10 0	7	10-8	6	4-12	438	24	3600	2	125	13-0	44-0	2692-1 *£7-10
Falcon	160	44	87-0	27-0	11-6	13-0	337	449	1666	Mr Pett	Woolwich	6-0-0	6	10-8	5	4-12	408	22	3600	2	125	7-10	40-0	2527-1 *£7-10
Sweepstakes	160	44	86-0	28-0	10-6	12-6	358	477	1665	Mr Edgar	Yarmouth	6-0-0	6	10-8	5	4-12	408	22	3600	2	125	7-10	40-0	2327-0- *£6-10
Guernsey	120	32	80-0	24-0	10-0	12-0	245	326	1654	Mr Shish	Walderwick	5-2-1	6	8-2	5	4-5	480	22	3200	2	85	5-3	34-0	1592-1
Success	130	36	86-0	26-0	10-6	12-6	309	412	1657	Capt Taylor	Chatham	5-4-0	6	8-16	5	4-7	480	22	3400	2	85	6-0	36-0	2297-1 *£7-10
Speedwell	110	28	76-0	24-0	10-0	11-0	232	309	1655	Mr Callis	Deptford	5-0-0	6	8-1	5	4-3	480	22	3000	2	80	5-0	30-0	1508-0- *£6-10
Dartmouth	110	30	80-0	24-9	10-4	12-0	260½	346⅚	1656	Mr Tippets	Portsmouth	5-2-1	6	8-4	5	4-2	480	22	3200	2	85	5-3	32-0	1693-5
Milford	110	30	78-0	24-8	10-0	13-0	252	336	1654	Mr Page	Wivenhoe	5-0-0	6	8-2	5	4-1	480	22	3000	2	85	5-3	32-0	1638-0
Forester	110	30	80-0	24-0	10-0	12-0	245	326	1657	Mr Furzer	Lidney	5-0-0	6	8-2	5	4-1	480	22	3000	2	85	5-3	32-0	1592-1
Garland	110	28	80-0	25-0	10-0	11-6	265	353	1654	Mr Furzer	Southampton	5-2-1	6	8-2	5	4-1	480	22	3000	2	85	5-3	30-0	1722-1
Mermaid	110	30	86-0	25-0	10-0	12-0	285	380	1651	Mr Graves	Limehouse	5-2-1	6	8-6	5	4-5	480	22	3200	2	85	5-3	32-0	1852-1
Nightingale	120	32	86-0	25-2	10-0	12-0	287	382	1651	Mr Bright	Horsleydown	5-2-1	6	8-6	5	4-5	480	22	3200	2	85	5-3	32-0	1865-1
Norwich	120	32	80-0	25-0	10-6	12-0	265	353	1655	Phi Pett	Chatham	5-2-1	6	8-6	5	4-5	480	22	3200	2	85	5-3	32-0	1722-1
Pearl	120	32	86-0	25-0	10-0	12-0	285	380	1651	Pet Pett	Ratcliffe	5-2-1	6	8-6	5	4-5	480	22	3400	2	85	5-3	32-0	1852-1
Eagle	110	30	85-6	25-8	10-0	12-0	297	396	1654	Capt Taylor	Wapping	5-2-1	6	8-5	5	4-5	480	22	3200	2	85	5-3	32-0	1930-1
Richmond	100	26	76-0	24-0	10-0	11-0	232	309	1654	Mr Tippets	Portsmouth	4-15-0	6	7-18	5	4-0	480	22	2800	2	80	5-3	30-0	1508-0
Little Victory	100	26	80-0	24-0	9-6	10-0	245	326	1666	Mr Lawrence	Chatham	4-15-0	6	8-1	5	4-4	480	22	3000	2	80	5-0	30-0	1592-1

RIGGING

Price of Cordage to Rig (£ s)	Price of Cables (£ s)	Price of Anchors (£ s d)	Price of Blocks, Tops, and Pumps (£ s)	Price of Sails (£ s d)	Price of Brass Guns (£ s)	Price of Iron Guns (£ s)	Price of Shot (£ s)	Price of Powder (£ s)	Price of Gunner's Stores New (£ s)	Price of Boatswain's Stores and Cordage (£ s)	Price of Carpenter's Stores (£ s)	Price of the Ship Completely Rigged and Stored (£ s d)	Charge of Victualling, 6 Months (£ s d)	Charge of Seamen, 6 Months (£ s d)	Charge of Officers, 6 Months (£ s d)	Charge of the Whole Ship Completely Rigged, Stored and Victualled and Wages for 6 Months (£ s d)
322-17	495-0	196-4	79-0	271-2	–	972-0	192-0	405-0	514-7	244-0	32-0	6866-3-2	1092-0-0	1121-5-0	432-15	9512-3-2
322-17	495-0	196-4	79-0	271-2	–	972-0	192-0	393-0	514-7	244-0	32-0	7090-3-2	1092-0-0	1121-5-0	432-15	9736-3-2
322-15	497-5	210-0	79-0	271-0	–	972-0	192-0	405-0	514-7	244-0	32-0	7905-11-8	1092-13-4	1121-5-0	432-15	10551-11-8
322-17	497-5	196-4	79-0	271-0	–	972-0	192-0	405-0	514-7	244-0	32-0	7060-3-2	1092-13-4	1121-5-0	432-15	9706-3-2
315-0	495-0	196-4	79-0	271-0	–	900-0	192-0	390-0	514-7	244-0	32-0	6501-11-2	970-13-4	971-15-0	432-15	8876-14-6
322-15	497-5	196-4	79-0	283-6	–	972-0	192-0	405-0	514-7	244-0	32-0	7321-19-8	1092-0-0	1121-5-0	432-15	9967-19-8
310-10	495-0	196-4	79-0	271-0	–	972-0	176-0	390-0	514-7	244-0	32-0	6592-1-2	970-13-4	971-15-0	432-15	8967-4-6
322-15	495-0	196-4	79-0	288-8	–	972-0	192-0	405-0	514-7	244-0	32-0	7673-5-8	1092-0-0	1121-5-0	432-15	10319-5-8
315-10	465-15	180-14	79-0	255-0	–	792-0	192-0	375-0	488-10	244-0	32-0	5401-19-0	1031-6-8	1046-10-0	432-15	7911-10-8
322-15	495-0	196-4	79-0	269-3	–	972-0	192-0	405-0	514-7	244-0	32-0	7673-11-4	1092-0-0	1121-5-0	432-15	10319-11-4
292-10	468-0	180-4	79-0	255-0	–	792-0	176-0	375-0	488-10	244-0	32-0	6075-4-0	970-13-4	971-15-0	432-15	8450-7-4
270-0	468-0	180-14	60-0	255-0	–	720-0	90-0	375-0	480-0	240-0	28-0	5694-4-4	970-13-4	1001-13-0	292-19	7959-9-4
270-0	468-0	180-14	60-0	255-0	–	720-0	90-0	375-0	480-0	240-0	28-0	5493-14-0	970-13-4	1001-13-0	292-19	7758-19-4
230-1	364-10	163-14	56-0	226-13	–	612-0	66-10	255-0	234-0	157-0	26-0	3981-18-4	728-0-0	702-13-0	292-19	5705-10-4
234-0	396-0	167-10	56-0	240-16	–	648-0	76-0	255-0	234-10	157-0	26-0	4788-6-8	788-13-4	777-8-0	292-19	6572-12-0
225-0	362-5	160-0	56-0	212-10	–	540-0	66-10	255-0	234-10	157-0	26-0	3787-15-8	667-6-8	627-18-0	292-19	5377-18-8
230-1	369-0	160-0	56-0	226-13	–	576-0	66-10	255-0	234-10	157-0	26-0	4049-19-4	667-6-8	627-18	292-19	5638-3-0
225-0	364-10	158-1	56-0	212-10	–	576-0	66-10	255-0	234-10	157-0	26-0	3969-1-0	667-6-8	627-18-0	292-19	5557-4-8
225-0	364-10	158-1	56-0	212-10	–	576-0	66-10	255-0	234-10	157-0	26-0	3923-11-0	667-6-8	627-18-0	292-19	5511-14-8
230-1	364-10	158-1	56-0	212-10	–	540-0	66-10	255-0	234-10	157-0	26-0	4022-12-0	667-6-8	627-18-0	292-19	5610-15-8
230-1	373-10	163-14	56-0	226-13	–	576-0	66-10	255-0	234-10	157-0	26-0	4217-8-4	667-6-8	627-18-0	292-19	5805-12-0
230-1	373-10	163-14	56-0	226-13	–	576-0	66-10	255-0	234-10	157-0	26-0	4230-8-4	728-0-0	702-13-0	292-19	5954-0-4
230-1	373-10	163-14	56-0	226-13	–	576-0	66-10	255-0	234-10	157-0	26-0	4087-8-4	728-0-0	702-13-0	292-19	5791-0-4
230-1	373-10	163-14	56-0	240-10	–	576-0	66-10	255-0	234-10	157-0	26-0	4231-11-8	728-0-0	702-13-0	292-19	5955-3-8
230-1	371-5	163-14	56-0	226-13	–	576-0	66-10	255-0	234-10	157-0	26-0	4293-3-4	667-6-8	627-18-0	292-19	5881-7-0
213-15	355-10	156-1	50-10	198-6	–	540-0	66-10	240-0	230-0	157-0	26-0	3471-2-8	606-13-4	553-3-0	292-19	5193-18-0
213-15	362-5	164-0	50-0	212-10	–	540-0	60-0	240-0	230-0	157-0	26-0	3848-0-0	606-13-4	553-3-0	292-19	5300-15-4

DEANE'S DOCTRINE OF NAVAL ARCHITECTURE, 1670

pp83-84

	Number of Men	Number of Guns	Length by the Keel (ft ins)	Breadth by the Beam (ft ins)	Depth in Hold (ft ins)	Draught of Water (ft ins)	Number of Tons	Tuns and Tunnage	When Built	By Whom Built	Where Built	Weight of Cordage to Rig (tons cwt lbs)	Number of Cables	Weight of Cables (tons cwt)	Number of Anchors	Weight of Anchors (tons cwt)	Number of Blocks and Deadeyes	Number of Sails	Yards of Canvas	Number of Boats	Barrels of Powder	Weight of Shot (tons cwt)	Weight of Guns (tons cwt)	Price of the Hull complete from the Builder (£ s d) (*cost per ton)
Drake	70	16	85-0	18-0	7-6	8-0	146	194	1653	Phi Pett	Deptford	2-12-0	4	4-16	5	1-3	408	22	1178	2	22	2-5	12-0	949-0- *£6-1
Hart	50	12	50-0	14-6	6-6	6-6	55	73	1657	Chr Pett	Woolwich	2-0-0	4	4-2	4	1-0	408	22	1000	2	10	2-0	8-0	357-1
Martin	70	16	64-0	19-4	7-6	8-0	125	166	1653	Mr Tippets	Portsmouth													
Roe Ketch	70	10	50-0	18-0	9-0	8-4	79	107	1665	Mr Page	Wivenhoe	2-0-0	4	3-6	4	0-15	170	19	1170	1	8	1-0	3-10	513-1
Truelove	60	16	60-0	18-0	7-4	7-8	103	137	1666	Capt Deane	Harwich	2-6-0	4	3-18	5	1-0	408	22	1180	2	22	2-0	12-10	669-1
Roebuck	80	20	64-0	19-6	9-2	8-4	136	181	1666	Capt Deane	Harwich	2-12-0	5	5-8	5	1-6	408	22	1250	2	22	2-5	13-0	884-0
Fanfan	35	4	44-0	12-0	5-8	6-0	33	44	1666	Capt Deane	Harwich	1-1-0	3	2-17	3	0-12	206	12	840	2	2	0-4	0-12	214-1
Francis	75	18	66-0	20-0	9-2	9-0	144	186	1666	Capt Deane	Harwich	2-16-0	5	5-8	5	1-6	408	22	1250	2	22	2-5	13-0	910-0
Spy	30	4	44-0	11-0	4-0	4-0	28	37	1666	Capt Deane	Harwich	1-2-0	3	2-14	3	0-10	180	12	800	2	2	0-4	0-12	182-0
Wivenhoe Ketch	60	10	50-0	17-6	9-0	8-4	79	105	1666	Mr Page	Wivenhoe	2-0-0	4	3-6	4	0-15	170	18	1170	1	8	1-0	3-10	513-1
Colchester Ketch	60	10	50-0	17-6	9-0	8-4	79	105	1666	Mr Allin	Colchester	2-0-0	4	3-6	4	0-16	170	18	1170	1	8	1-0	3-10	513-1
Katherine Yacht	20	12	50-0	18-4	7-9	7-2	94	125	1660	Pet Pett	Deptford	2-3-0	4	3-6	4	0-15	98	14	1030	1	5	0-10	2-14	658-0 *£7
Ann Yacht	20	12	51-0	18-0	7-9	7-0	91	121	1660	Chr Pett	Woolwich	2-3-0	4	3-6	4	0-15	98	14	1030	1	5	0-10	3-4	637-0
Henrietta	20	12	52-0	18-10	8-0	7-1	108	147	1663	Chr Pett	Woolwich	2-3-0	4	3-8	4	0-16	98	14	1040	1	5	0-10	3-2	686-0 *£7
Monmouth	16	10	48-0	18-0	7-6	7-2	86	114	1665	Mr Castle	Rederiffe	2-0-0	4	3-2	4	0-15	98	14	1010	1	4	0-8	2-0	602-0 *£7
Mary Yacht	20	10	50-0	18-6	7-4	7-0	92	122	1660	The States	Holland	2-3-0	4	3-8	4	0-17	98	14	1020	1	4	0-8	2-0	644-0
Bezaan Yacht	10	6	40-0	16-0	6-0	5-6	54	72	1663	The States	Holland	1-4-0	3	2-7	3	0-12	56	10	810	1	2	0-5	0-15	540-
Kitchen	12	4	48-0	17-6	8-0	8-0	76	101	1652	Mr Page	Wivenhoe	2-3-0	4	3-6	4	0-15	98	14	1010	1	1	0-4	0-11	400-
Deptford Ketch	60	12	52-0	19-0	9-4	8-4	99	132	1656	Mr Shish	Deptford	2-0-0	4	3-8	4	0-16	170	18	1170	1	9	1-0	3-10	581-1 *£6-
Merlin Yacht	20	8	52-0	18-10	8-0	7-1	108	147	1666	Mr Shish	Jamaica	2-3-0	4	3-8	4	0-16	98	14	1040	1	5	0-10	3-2	686- *£7
Saudadoes	20	8	52-0	17-6	9-0	8-6	87	116	1669	Capt Deane	Portsmouth	2-5-0	4	3-12	4	1-3	408	18	1050	1	5	0-10	3-14	609-

RIGGING

Price of Cordage to Rig (£ s)	Price of Cables (£ s)	Price of Anchors (£ s d)	Price of Blocks, Tops, and Pumps (£ s)	Price of Sails (£ s d)	Price of Brass Guns (£ s)	Price of Iron Guns (£ s)	Price of Shot (£ s)	Price of Powder (£ s)	Price of Gunner's Stores New (£ s)	Price of Boatswain's Stores and Cordage (£ s)	Price of Carpenter's Stores (£ s)	Price of the Ship Completely Rigged and Stored (£ s d)	Charge of Victualling, 6 Months (£ s d)	Charge of Seamen, 6 Months (£ s d)	Charge of Officers, 6 Months (£ s d)	Charge of the Whole Ship Completely Rigged, Stored and Victualled and Wages for 6 Months (£ s d)
18-13	215-0	39-8-0	42-0	83-8-10	–	216-0	29-0	66-0	125-16	89-0	17-0	1991-5-10	424-13-4	244-15-0	418-12	3079-6-2
90-0	184-0	34-0-0	40-0	70-16-8	–	144-0	24-10	30-0	100-0	80-0	17-0	1172-6-8	303-6-8	244-15-0	269-2	1989-10-4
90-0	148-10	25-5-0	23-0	82-17-6	–	63-0	11-0	24-0	60-0	45-0	14-0	1100-7-6	404-13-4	145-10-0	373-15	2044-5-10
103-10	175-10	34-0-0	42-0	83-1-8	–	216-0	24-10	66-0	125-16	89-0	15-0	1636-7-8	364-0-0	244-15-0	343-17	2588-19-8
18-13	243-0	40-16-0	42-0	88-10-0	–	234-0	29-0	66-0	125-16	89-0	17-0	1980-15-0	485-13-4	244-15-0	493-7	3204-4-6
67-10	128-5	19-18-0	7-0	59-10-0	–	10-12	2-5	6-0	16-0	20-0	8-0	559-9-10	212-6-8	124-9-0	224-5	1120-10-6
26-0	243-0	44-16-0	42-0	88-10-10	–	234-0	29-0	66-0	125-16	89-0	17-0	2015-2-10	455-0-0	244-15-0	455-19	3150-17-4
44-10	121-10	17-0-0	7-0	56-13-4	–	10-12	2-5	6-0	16-0	20-0	8-0	496-9-4	182-0-0	124-9-0	186-15	989-15-4
90-0	148-10	25-10-0	23-0	82-17-6	–	63-0	11-0	24-0	60-0	45-0	14-0	1100-7-6	364-0-0	145-10-0	373-15	1983-12-6
90-0	148-10	27-4-0	23-0	82-17-6	–	63-0	11-0	24-0	60-0	45-0	14-0	1102-1-6	364-0-0	145-10-0	373-15	1985-6-6
96-15	148-10	25-10-0	30-0	72-19-2	405-0	–	15-10	15-0	60-0	40-0	14-0	1571-4-2	121-6-8	124-9-0	97-3	1914-3-4
96-15	148-10	25-10-0	30-0	72-19-2	480-0	–	15-10	15-0	60-0	40-0	14-0	1625-4-2	121-6-8	124-9-0	97-3	1968-3-4
96-15	153-0	27-4-2	30-0	73-13-4	465-0	–	5-10	15-0	60-0	40-0	14-0	1666-2-4	121-6-8	124-9-0	97-3	2009-1-6
90-0	139-10	25-10-0	30-0	71-10-0	300-0	–	4-8	12-0	60-0	40-0	14-0	1388-18-10	97-1-4	124-9-0	61-5	1677-14-8
96-10	153-0	28-18-0	30-0	72-7-0	300-0	–	4-8	12-0	60-0	40-0	14-0	1455-6-0	121-6-8	124-9-0	97-3	1798-5-2
54-0	105-15	19-0-0	15-0	57-6-6	112-10	–	3-6	6-0	20-0	10-0	5-0	948-14-6	60-13-8	94-5-0	37-7	1141-1-4
96-15	148-10	25-10-0	23-0	71-10-0	–	9-18	2-5	3-0	18-0	20-0	12-0	830-7-10	72-16-0	94-5-0	52-6	1049-16-4
90-0	153-0	27-4-0	23-0	82-17-6	–	63-0	11-0	27-0	60-0	45-0	14-0	1177-16-6	364-0-8	145-10-0	373-15	2061-1-6
96-15	153-0	27-4-0	30-0	73-13-4	465-0	–	15-10	15-0	60-0	40-0	14-0	1666-2-4	121-6-8	124-9-0	97-3	2009-1-6
101-5	163-0	40-5-0	40-0	74-7-6	555-0	–	5-10	15-0	60-0	40-0	14-0	1716-7-6	121-6-8	124-9-0	97-3	2059-6-2

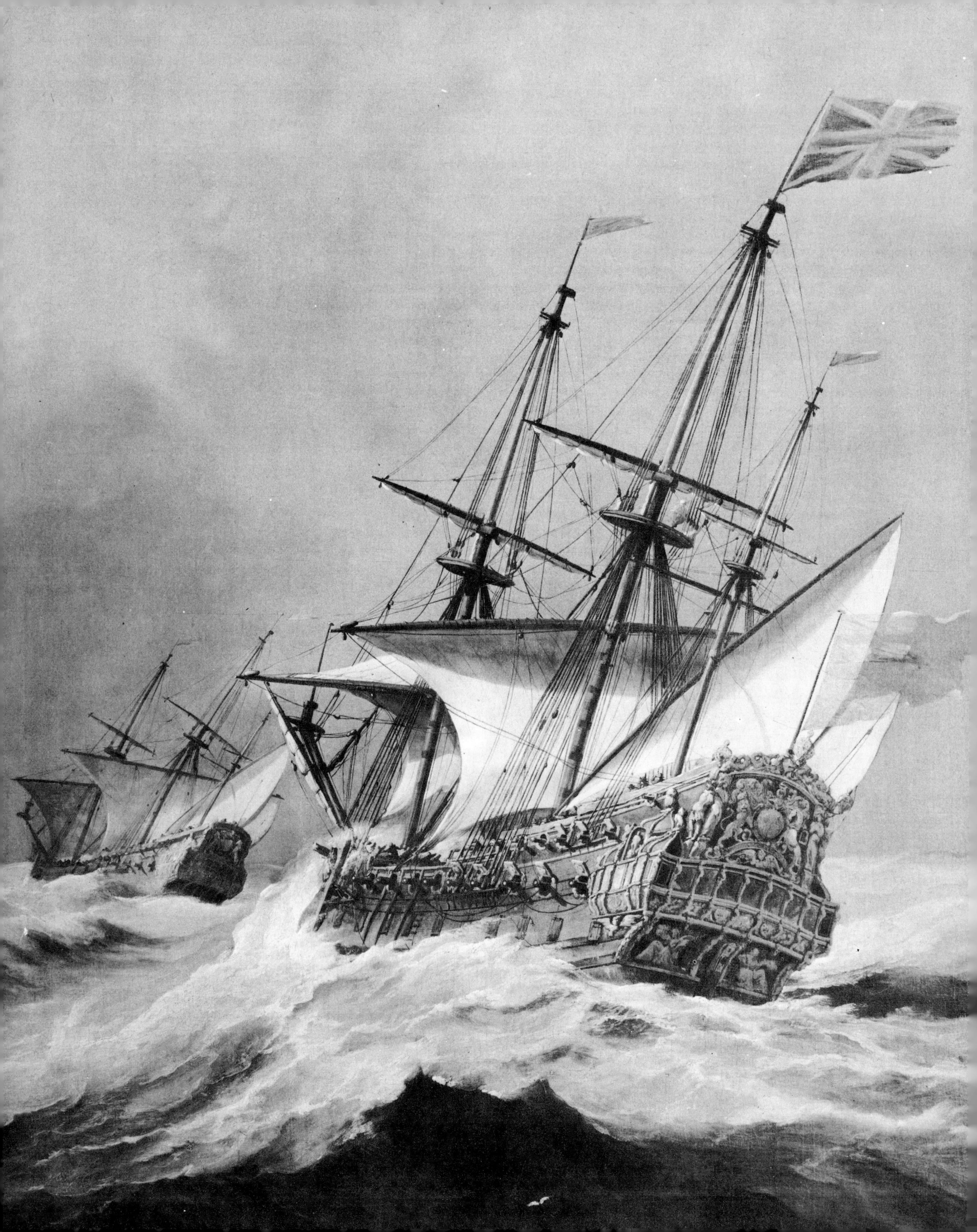

APPENDIX

The first purpose of this section is to fill some of the gaps in Deane's *Doctrine*, with the use of contemporary documents and drawings. Deane says very little about the structure of ships, the arrangement of their upper decks, or their decorations, and his *Doctrine* needs to be supplemented to give a complete account of naval shipbuilding of the time. Most of the documents are previously unpublished, and date from within 10 years of 1670. The second purpose is to use this information to show what one of his ships would have looked like, and for this I have chosen the *Resolution*, for reasons which are outlined later.

SOURCES FOR APPENDIX

Specification for 3rd Rate: Public Record Office, SP 46/136 f227. Details of the *Warspite* as built: British Library, Additional Manuscripts, 9304, f206. Contract for carved works: Public Record Office, SP 29/185 f49. Section of the *Britannia:* Coronelli's *Gli Argonauti*, 1691. Sections of the *Rupert*: Pepys Library, 2501. Cross sections and deck plan of 1st Rate: Pepys Library, 2934. Drawing and painting of the *Resolution*: National Maritime Museum. Sails of 1st Rate: British Museum Additional Manuscripts, 9303, f19.

Left: Van de Velde's famous painting of the *Resolution* in a gale. *(National Maritime Museum)*

The dimensions and scantlings of all timbers for the building of a 3rd Rate ship for the King.

The parts *in italics* are those which were added when the specification was amended. The figures given in parentheses refer to the original text.

A CONTRACT SPECIFICATION

This specification was probably used for the six 3rd Rates of 1666-7, especially for the two contract-built ships, *Warspite* and *Defiance*, for merchant builders were generally given detailed specifications, whereas it was felt that the dockyards could be trusted to decide the details for themselves. It is probable that the original, before amendment, is the specification for the *Speaker* of 1649 (renamed *Mary* in 1660), since her main dimensions correspond with those given, and she was designed by Christopher Pett, who drew up the first version of this specification.

Length of the keel, 116 feet. Breadth from outside to outside of plank, *36 feet 0* (34 feet). Depth in hold to the upper edge of the beam *15 feet* (14 feet). Breadth at transom, *24 feet* (21 feet). Rake fore and aft, *33 feet* (24 feet). The keel to be 15 inches square in the midships, and 11 inches thwartships (at least) abaft. The stem to be 14 inches thwartships and 15 up and down. The scarfs of the keel to be 4 feet in length and coaked into one another and well bolted with 6 bolts of an inch in each scarf, and to bring on a false keel of 4 inches, to be well fastened with spikes. To have a good substantial false stem of *10 inches* (8 inches) thick and *2½ feet* (2 feet) broad, to be well bolted with *1¼ inch* (blank) bolts into the stem. The stem post to be 2 feet fore and aft alow, and to have a substantial knee, to be well bolted with an inch and a quarter bolt through the keel and post. And to work the run up with long timber, to be well bolted into the keel with an inch and a quarter bolt. The floor timber to be in length *24 feet* (22 feet), up and down one foot, and fore and aft 14 inches. Room and space to be 2 feet 4 inches, the dead rising to be 4 inches at least. The lower futtocks to fill up the room, and to have 6 or 7 foot scarf in the midships. The other tier of futtocks to have 6 feet scarf at least, and the navel timber to be 10 inches in and out at the breadth. To have a substantial kelson of 16 inches up and down and 18 inches broad, to run fore and aft, to be well bolted with an inch and quarter auger in every other timber, and the rest of the timber under to be well bolted with an inch and a quarter auger before the said kelson be brought on. To have 6 strakes of footwaling, three of 8 inches thick, two of 7 inches, and one of 6 inches, each strake to work 16 inches broad. To have two footwales of middle bands, to run fore and aft, to be 6 inches thick and 16 inches broad, to be tabled with an inch and a half into each other. To have two strakes of 7 inch clamps, to run fore and aft, of 16 inches broad, to be tabled 2 inches into each other with a flemish scarf of hook and butt, to keep her from writhing. To berth up all the work between with 4 inch plank, only to leave an opening to air the timbers under the clamps. To fay all scaling boards fore and aft as is necessary. To have 5 bends of riders, to be well bolted with an inch and a quarter auger, the floor riders to have eight bolts in each, and the futtock riders 6 apiece. To place *ten* (eight) half beams in hold, of a foot up and down and *17 inches* (14 inches) fore and aft, to be doubled, kneed with knees at each end, with four bolts in each knee of an inch and a half quarter auger. To lay the said beams 5 feet from the lowest edge of the beams of the gun deck. To lay carlines some 8 feet from the side from beam to beam in the wake of the hatchway, and to place ledges from the side to the said carlines, and to lay the seams with deal or board, a convenient length from the railing of the cables. To lay platforms for the cook room and all store rooms in hold, and to make eight cabins in a platform abaft for the lodging of the *officers* (men), and to fit the steward's room and carpenter's store room, and to make all such conveniences as shall be requisite. In hold to fay a step for the foremast, to be well bolted with eight bolts of an inch and a quarter auger. To fay a saddle some 6 feet long and 2 feet broad at least, for the step of the mainmast. To place four breast hooks under the hawse and one above, of 12 inches up and down, to have good long arms and to be well bolted with 6 or 7 bolts in each hook, of an inch and a quarter auger. To have three transoms below the gunroom ports, to be a foot up and down, and to be *18 inches* (2 feet) asunder, with a chord at the lower end of the fashion piece, the said transoms to be well kneed with a knee at each end, to be well bolted with an inch bolt, 6 bolts in each knee. The beams of the gun deck to be 14 inches up and down and 16 inches fore and aft. To lay one beam under every port and one between, to be well kneed with 4 knees to each beam, the one fore and aft, the other up and down, and to bolt the said knees with five bolts to each knee, of an inch and a half quarter auger. To lay two tiers of carlines on the side of the said deck fore and aft, to be 8 inches deep and 10 inches broad, the ledges to lie within 10 inches one of another, and to be in substance 4 inches up and down and 5 fore and aft. To put in 5 pair of cross pillars, to be 9 inches athwartships and 10 fore and aft, to be well fastened and stove into the beams of the gun deck, half beams, and the heads of the floor riders, with a bolt of an inch in each said beam and rider, and to have a small knee at the foot of each pillar, to be bolted with two bolts of an inch in each arm of the said knee. To fay standing pillars of 7 inches square under every other beam. To bring on waterways of 6 inches thick on the gun deck, and some 15 or 16 inches broad, and to lay the said deck with good 3 inch plank. To place two pair of bitts with cross pieces on the gun deck, the one to be 18 inches, the other 16, inches square. To have four substantial standards placed to them, with three bolts to each standard of an inch and a half quarter bolt. To place four substantial hawse pieces and to cut out four hawse holes in them. To lay coamings of a convenient height and length in the wake of the main hatchway, and to make hatches to the same, to be laid with two inch plank and fitted with iron *rings,* (blank), hinges to them, and to make all other conveniences of hatches, gratings, and scuttles for the stowing of provisions and for the conveniences of steward's room, lodging for men, cook room, boatswain's and gunner's store rooms. To lay all partners for mainmast, foremast and mizzen mast, the main capstan on the said deck. To have an iron hoop on the head, and to fit two iron paules to the same, with capstan bars, and likewise to fay a sole for the bow capstan. To bring on two strakes of spirkett rising, of 4 inch plank, fore and aft of 14 inches broad, to be wrought with a flemish scarf and to be tabled into each other. To cut out 13 ports on each side of the said deck, and two aft, and to put in port cills to them. To bring on a rising of 4 inch plank under the beams of the upper deck, to have a flemish scarf of hook and butt, and to shut up all the work between with good 3 inch plank. To fay four pairs of substantial standards with shoulders of 2 inch plank under them, and to bolt them with 6 bolts in each, with an inch and a half quarter auger. To bring on a transom over the gunroom ports of 9 inches thick, to be kneed with a knee at each end, and to be bolted with 5 bolts in each knee, of an inch auger. *To put in a pair of standards upon the transom over the helm port to run up to the rising of the great cabin* (blank). To make a manger and to place turned pillars fore and aft under the upper deck beams where need shall require. The beams of the upper deck to be 12 inches fore and aft and 10 up and down, to lie *5 and 6* (6 and 8) feet asunder *except in the hatchway* (blank). To knee the said beams with two knees at each end, one fore and aft, the other up and down, and to bolted with four and 5 bolts in each knee, of an inch *to be 6 feet, and eight knees betwixt plank and plank between decks* (blank). To fay a transom abaft, to lie even with the beams, to be kneed at each end and bolted with four bolts in each knee, the said knees to be placed fore and aft. To have two rows of fir carlines fore and aft with substantial ledges, the carlines to be 6 inches up and down and 9 inches broad. To have long carlines in the waist of 9 inches square, to make the coaming for the ledging of the gratings. To bring on waterways of 4 inches broad, and to lay four strakes of 2 inch plank in the wake of the guns fore and aft. The rest of the said deck to be laid with spruce deals. To lay all coamings and head ledges, and to make all such gratings fore and aft as shall be needful for the venting of the smoke of the ordnance. To fay one knight topsail sheet bitt, and jeers for the masts to be made, and place a jeer capstan with partners and bars for the same, to have an iron hoop on the head of it, and to fit iron paules with bolts to it, and to fay all partners on the said deck for the masts. To cut out a tier of ports fore and aft, and to fay port cills to them. To bring a strake of spirkett rising of 3 inches fore and aft, of 17 inches breadth, with a flemish scarf of hook and butt, and to fay a gunwale of 3 inch plank in the waist, to serve for the lower cill of the ports. To make a large quarter deck and forecastle, to have the beams of both to be 5 inches up and down and 6 fore and aft, to lie two feet asunder, and to knee every other beam with one knee at each end, to be bolted with four bolts of 3 quarter inch auger in each knee. To bring on two pieces of rising of 7 inches square for to lay the ends of the beams of the great cabin, on which must be dovetailed into the said piece and bolted at each end with a bolt of ¾ inches. The said beams to be 4 inches up and down and 6 inches fore

Below and opposite: Sections of a ship, probably the *Rupert*, from Deane's 'Method of Measuring the Body of a Ship'. *(Pepysian Library)*

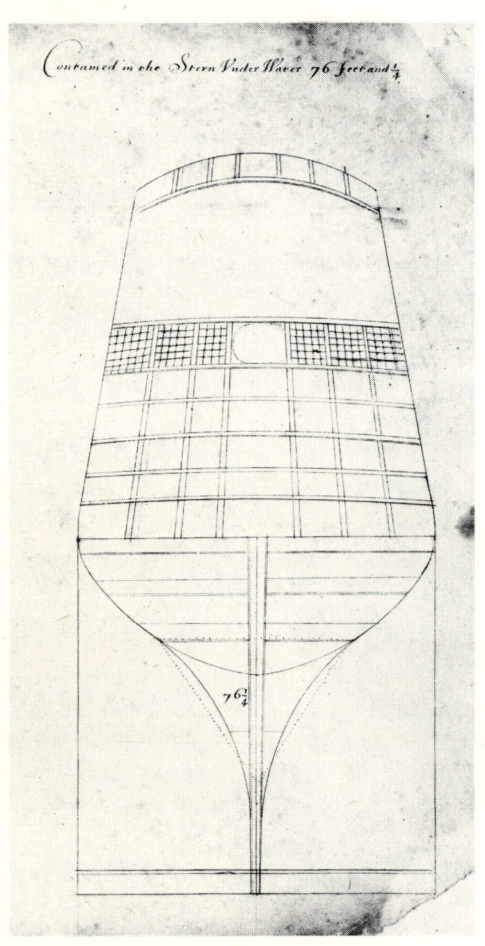

and aft, and to lie 2 feet 4 inches asunder. To bring on a rising, of spruce deals of two inch plank under the beams of the quarter deck and shut up all the rest of the work with good ready deals. To bring on a good rising of 3 inch plank under the beams of the forecastle and to lay the said beams 2 feet asunder, and to shut up the rest of the work between with 2 inch plank. To knee every other beam with a knee up and down at each end, to be bolted with four bolts in each knee of three quarter inch. To lay the said forecastle with good spruce deals and to have a wing board of 3 inch plank, to be pricked over to the outside plank. To fay coamings and head ledges and to make a grating for the venting of the smoke of the guns on the said forecastle. To fay two substantial catts, to be well bolted with four bolts in each, of an inch. To fay supporters to them, to be well bolted with three bolts of an inch in each of the said supporters, to be carved, and to cut out two sheaves in each cathead. To make three bulkheads, two to the forecastle and one to the quarter deck, to have good substantial stanchions, and have the 3 inch plank thwartships for the feet of the timbers to stand in, and to berth the said bulkheads up with spruce deals. To make all doors, and to cut out all ports in each bulkhead, and to hang the said ports and doors with substantial hinges. To lay a beam (to round forward) for the fore peak, to lie even with the lower edge of the ports in the forecastle, to be kneed with a knee at each end and bolted with four bolts in each knee, of ¾ inch, and to lay the said forepeak with 2 inch plank. To put in four 5 inch scuppers, two in the manger, and two for the pumps, and eighteen 4 inch scuppers for the lower deck. To put in ten 3 inch scuppers in the waist and forecastle, and six two inch under the quarter deck. To fit all ring bolts and eye bolts to all ports on the lower and upper deck, four bolts to each port, them of the lower deck to be an inch and a quarter bolt, and the upper ones to be an inch bolt. To frame a staircase under the quarter deck, and to make a ladder to go down to the gun deck, and another to go down in the waist, and to place turned pillars under the beams of the quarter deck, and to fit a long table and benches. To make gunner's and purser's cabins on the gun deck, and a bulkhead (thwartships) to enclose the gunroom. To make two standing cabins a side under the quarter deck, and two of a side in the forecastle, and one of a side before the bulkhead of the quarter deck. To make a large great cabin with a partition in the middle, and to make all doors, and cut out all lights belonging to the same. To bring on a standard one a side before the bulkhead of the great cabin, on the upper deck, with one long arm to run up the side, to be well bolted with five bolts of an inch in each standard. To cut out two round ports abaft, and to fay 2 transoms, the one under the lights of the great cabin, the other even with the beams of the great cabin, to be kneed with one knee at each end, fore and aft, to be bolted with four bolts of ¾ inch in each knee. To make a companion at the after end of the quarter deck, to lie a convenient height for the cutting out of the lights in it, to have the better sight of the sails, and likewise to afford length for the whipstaff. To make a large roundhouse and to lay the same with good spruce deals, the beams to lie 2 feet asunder, and to have every other beam with a knee at each end, to be bolted with four bolts in each knee, of half an inch. To make a bulkhead at the foremost part of the roundhouse. To make doors and ports in the said bulkhead, the same to be berthed up with ordinary deal, landed and inbowed. To make four cabins in the roundhouse and to cut out a round port in each side in the cuddy. To fit the same with ring bolts and eye bolts. To make such windows and scuttles to the said cabins as may be needful, and to put one two inch scupper of a side in the said cuddy. To fay a transom, to lie even with the beams of the roundhouse, to be kneed with a knee at each end and bolted with four bolts in each knee of half an inch, the knees to be placed fore and aft. To make a ladder on each side, to be fitted with pillars and rails, to go up on the quarter deck. To make two small cabins afore the bulkhead of the roundhouse, and two more upon the poop. To fit the bulkhead with rails and gunwales, and berthed where need requires. To plank the said ship from the keel to the lower edge of the ports at the gun deck with 4 inch plank, and to have two firm wales of 13 inches up and down and 8 in and out, and a chainwale of 10 inches up and down and 6 in and out. To work 3 inch plank as high as the lower edge of the ports of the upper deck. The rise of the quickwork of the waist and forecastle to be 2 inch plank, and the quarter deck to be spruce deals. To make and hang all ports fore and aft with hoops and hinges, and to fit the said lids with tackles and rings. To make a fair head, with a firm substantial knee, to be well bolted with an inch and a quarter bolts, and cheeks treble rails with trailboard, fair and kneed. Beast and bracket, kelson, and make points and to fay linings for the hawse. To have a fair lower and upper counter, with rails, badges and brackets. To have a fair pair of galleries, with rails, carved badges, and brackets with lights and casements, and the turrets of them to be plated with white plates twice over. To have a fair upright with rails and brackets, windows and casements into the great cabin, with a complete pair of carved arms, with pilasters, and side forms with a fair taffrail aloft. To bring on a substantial gripe, to be well fastened with an inch and a quarter bolts with iron dovetails alow, and to have an iron stirrup in the sling, well bolted. To fay main, fore and mizzen channels, to be well bolted, with chain plates and bolts, as many as shall be requisite. To make and hang a substantial rudder, with bands, gudgeons, and painters. To have a mask head carved at the upper end and to fit a tiller, whipstaff and rail to the same. To rail, gunwale and plank sheer and garnish the said ship with brackets, and having pieces complete, and to fay chesstrees, sheets, blocks, kevels and ranges, as many as need shall require, and to make plates and other necessaries for the rigging of the said ship. For the painting, gilding the said ship where it required, and to make a complete set of masts. *To fit all joiners work, carving and painting and gilding, with suit of masts and yards complete.*

(signed) Christopher Pett
Woolwich, 22nd November 1664

(written on back) Copy, November 22nd, 1664
Dimensions for a 3rd Rate ship with all circumstances for a contract for building one. First sent us by Mr Chr Pett, then corrected by Sir J Mennes and Sir W Batten, and lastly perfected by Commissioner Pett.

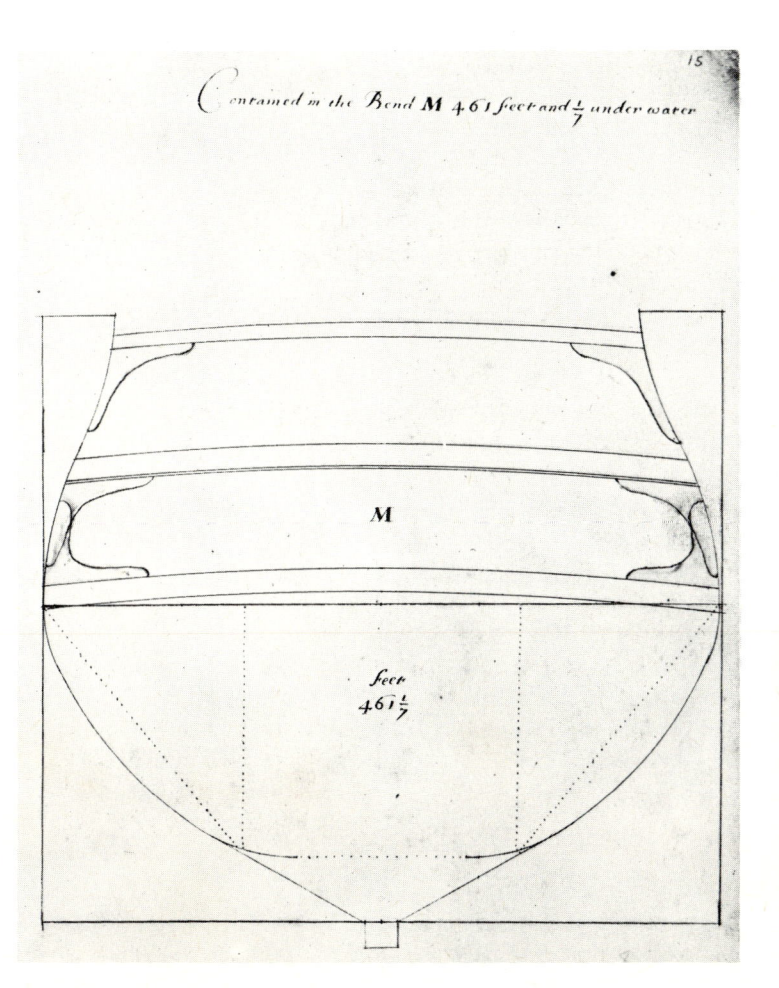

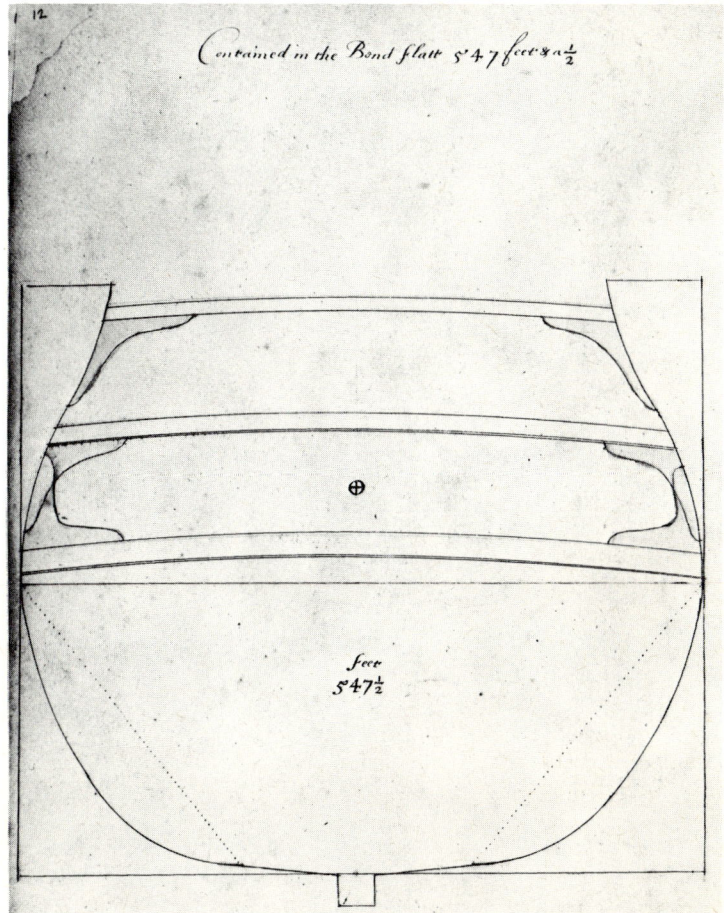

REPORT ON THE *WARSPITE*

The two ships built by contract to the preceding specification were rather larger than the contract demanded, partly because the King had instructed the builders to increase the breadth. The actual increase in size was paid for, for builders were normally paid according to the actual dimensions of the ship as completed, but the builders had also decided that some of the scantlings had to be increased to make the ship stronger. This is the report to the Navy Board on the extra work done to the *Warspite*. She was built by Henry Johnson in the Blackwall shipyard, and was launched in 1666. The comments and prices in the margin were apparently added later by the Navy Board.

Report on the Warspite.

Right Honourable,

According to your warrant to us directed, we have made our repair on board His Majesty's Ship the *Warspite*, and have strictly compared the performance of the said ship with the builder's contract, and do find all things performed according to the tenor thereof, both in respect of materials and workmanship. We also find several more works to be done by them more than contract, the particulars of which are as followeth:

One strake of middlebands of 6 inches thick and 16 inches broad, they being obliged to but 4 inch plank.	This is not more than contract.
One transom and a pair of knees, with all bolts to the same, being 13 in number.	10.0.0
One strake of four inch plank on each side the hatchway of the lower deck, which is bolted into every beam, they being to make a three inch without bolts.	08.0.0
They have fitted the cisterns and cases about the pumps with the dales for the passage of the water through the side more than contract.	This charge might well have been saved, for her ordnance lieth high enough above water.
For a strake of three inch plank wrought about the bows on the strake of spirketting from the stem to the after part of the bulkhead, being an inch thicker than contract.	There was no need of this.
For planking all the rest of the waist with two inch plank from the strake of the spirketting to the top of the waist, at which place is a four inch plank let into the timbers, and filled up between the heads of the said timbers to make all tight work on both sides — they being only to shut up to the lower edge of the ports by contract.	5.0.0
A strake of rising of three inch plank brought under the beams of the quarter deck in lieu two inches.	2.0.0
For shutting all the ceiling on both sides under the quarter deck with two inch plank, in lieu of ordinary deals mentioned in contract.	4.0.0
For bring on one strake of spirketting with gunwales on both sides upon the forecastle, more than the contract.	No need of that.
The same also upon the quarter deck more than contract.	
For the building of a coach on the quarter deck with carpenters', joiners', and painters' works and all iron works to the same, more than contract.	50.0.0
For one channel brought on fore and aft of five inches and ten inches broad, being two inches thick more than contract.	35.0.0
Total	104.0.0

All which we leave unto your honour's command and remaining

Woolwich, 30th of June 1666.

(Signed) Christopher Pett
Jonas Shish.

APPENDIX

A MANIFEST FOR CARVED WORK

This list, which is headed 'account of particular works are to be done to a new ship at Harwich by James Christmas, carver, for which he is to have payment', is dated February 1665, and was apparently used for the *Resolution*.

Bow view of the *Resolution*. (National Maritime Museum)

Carved work for the 3rd Rate ship at Harwich.

A lion for the head, with a trail board and sixteen brackets, and a frieze under the upper rail. The after end of the upper rail carved. Two supporters and two faces for the catheads. A head for the stem. Two faces for the cable holes. Six seams and brackets for the bulkhead of the forecastle. Two ports, six drifts or hancing pieces. Two chesstrees for the sails.

Six window pieces for the cabin and coach. Four lower counter pieces. Six masked heads, and a fair one for the rudder head. Four badges for the second counter, and four ports for the same, with sixteen brackets. A piece of King's arms. Two pilasters — one each side.

One taffrail for the top of stern

Two quarter pieces, one each side thwart.

Two figures for the quarters.

Thirty-six ports for the upper guns.

Ten brackets and two ports for the steerage bulkhead, and a piece over it.

Ten brackets for the forecastle bulkhead, and two ports.

Eight brackets for the coach bulkhead, and two ports.

The belfry afore.

One well for the great cabin, and thirty brass heads and a frieze.

For the staircase, eight brackets and buttons, with two badges.

For the well in the coach, two knight heads, with the timber head at the bulkheads.

121

The 3rd Rate Resolution, 1667

The *Resolution* of 1667 was possibly Deane's most successful ship. She was begun at Harwich Dockyard early in 1665, and launched in December 1667. She was too late for the Second Dutch War, so she was taken to Portsmouth and laid up. In March 1669 she began fitting out as Sir Thomas Allin's flagship for an expedition against the Barbary corsairs. The expedition was not a great success, since even the best English ships were too slow to catch the Mediterranean ships, and the squadron returned home at the end of 1670. On the way back it encountered a gale off Lisbon, which is shown in Van de Velde's famous painting. The *Resolution* was paid off in January 1671. A year later she was put into service again, and in March 1672 she took part in the unsuccessful attack on the Dutch Smyrna convoy which opened the Third Dutch War. The following month she formed part of the Red squadron at the Battle of Solebay. She was heavily damaged in the fighting, and was 'much disabled in her masts, yards, and sails'. She was repaired at Sheerness under the supervision of Deane, and set sail again in June. In September she cruised off Yarmouth, when she captured at least 23 Dutch merchant ships. After a refit, she was sent to cruise off the Scilly Isles, and she captured a privateer of 30 guns. She rejoined the main fleet in May 1673, and took part in the First Battle of Schoonveld, being 'much wounded', and in the Battle of the Texel.

Like most ships she was little used in the later years of Charles II's reign, and fell into disrepair, though not as much as some. She was repaired by the Special Commission of 1686-8 at a cost of £1292, and was the flagship of the fleet under the Earl of Dartmouth which failed to stop William of Orange from landing in 1688. In 1689 she escorted a convoy to the Mediterranean, and in 1691 she was present at the Battle of Barfleur. In 1692 she was made flagship for Admiral Wheeler's squadron, being chosen in preference to the *Rupert*, 'an old and rotten ship'. The squadron failed to capture Martinique and Newfoundland, and returned to

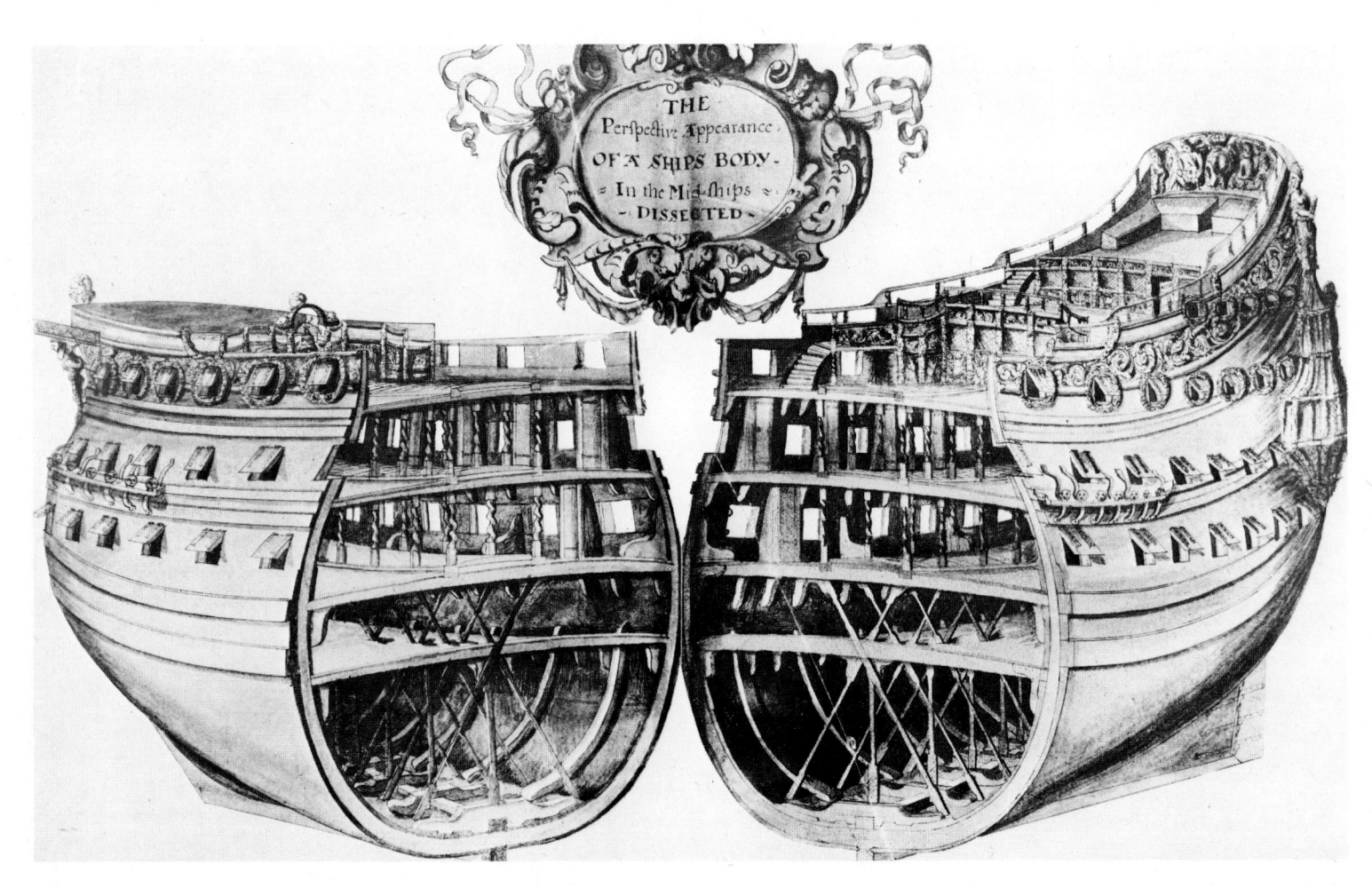

Cross sections of a 1st Rate, c1680. *(Science Museum)*

England in October 1693. In May 1694, with the *Monmouth,* she drove ashore and burnt one French frigate and 35 merchant ships. In 1695 she took part in the blockade of Dunkirk.

After several years of almost continuous service, she was found to be in poor condition, and in February 1697 she was put in dock for a rebuild, which was completed in April 1698. She was immediately fitted out for service in the Channel, but was soon found to be leaking. The poor quality of her repairs caused a scandal, which led to changes in the system of rebuilding.

In the Great Storm of November 1703 she was lost off the Sussex coast.

These drawings of the *Resolution* of 1667 are based on the lines given in Deane's *Doctrine,* on the dimensions given in tables on page 73 of that work, on two Van de Velde drawings and one painting, and on the contract for the carved works. It is impossible to be sure of every detail, particularly since the different sources are to a certain extent in conflict, and the details, especially of the decoration, are often rather vague. Donald McNarry has made a model of the *Resolution* based on the same sources, and in some ways his interpretation differs from mine, especially in the shape of the quarter galleries. Nevertheless I feel that this is as close as we are likely to get to a complete drawing of a ship of this period, except where a detailed contemporary model is available.

Sir Wescott Abell, in his book *The Shipwright's Trade,* suggests that the 3rd Rate drawn in the *Doctrine* is close to the *Rupert* of 1666, but there are several reasons to doubt this. Firstly the drawings are not of a specific ship, since they show 14 gun ports instead of 13 on the lower deck, and have decorations which were out of date by 1666. Secondly, the ship in the draughts is as close to the *Resolution* as the *Rupert*. The *Rupert* was slightly shorter on the keel, the *Resolution* slightly longer. Both were slightly broader than Deane's 36 feet, the *Rupert* at 36 feet 3 inches, the *Resolution* at 37 feet 2 inches but although Deane suggests that the 36 feet is measured outside the plank, it is clear from what follows that it is measured inside. In that case the real breadth should be 36 feet 8 inches, allowing for 4 inches of plank on each side. Finally, the *Resolution* was Deane's last 3rd Rate before writing the *Doctrine,* and there is no evidence that anything had happened to change his style between 1667 and 1670.

The drawings were traced for publication by John Roberts.

Longitudinal section of the *Britannia* of 1682. *(National Maritime Museum)*

DEANE'S DOCTRINE OF NAVAL ARCHITECTURE, 1670

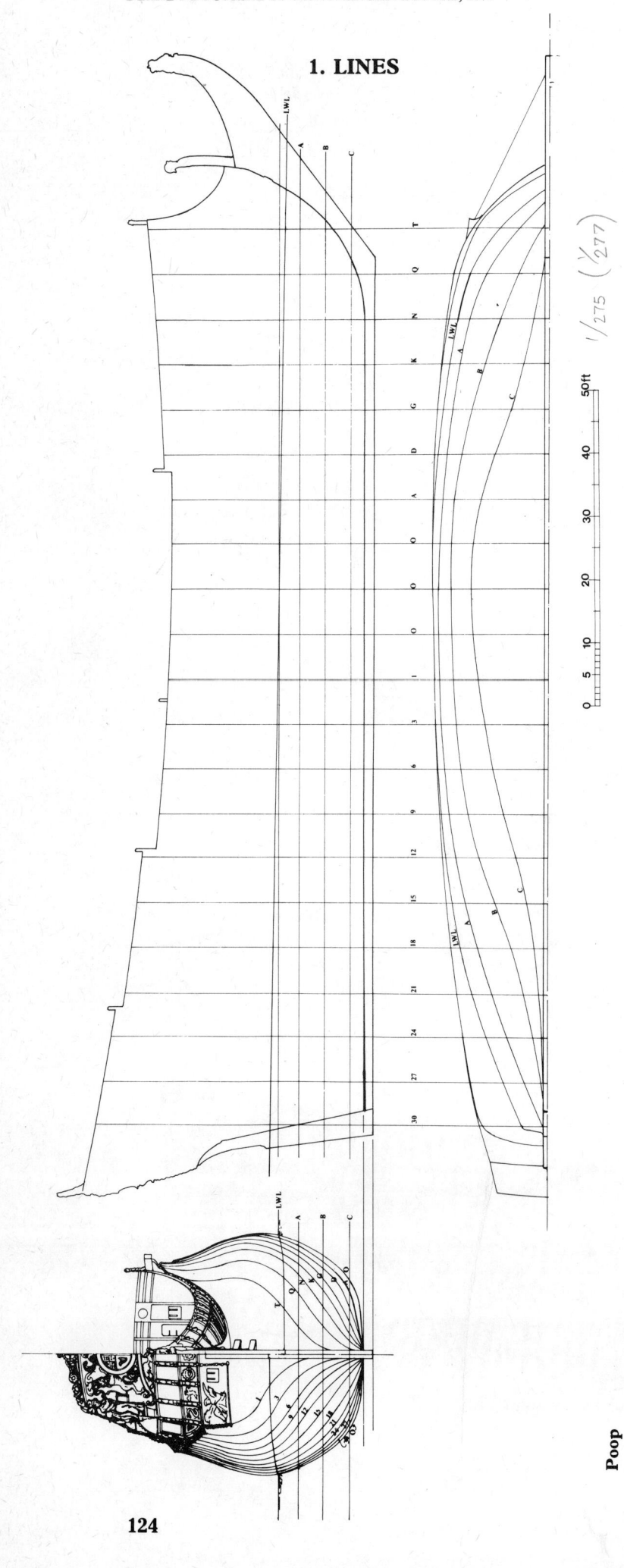

1. LINES

2. DECKS

Deane gives very little information on deck arrangements except in his longitudinal section and plan of gun deck. Cabin arrangement is largely based on the establishment of 1673 *(Catalogue of Pepysian Manuscripts,* Vol. I, pp189-192) with some reference to Dummer's deck plans of a 1st Rate in the Pepysian Library. It has also been compared with a rather later set of plans, those of the new *Resolution* of 1708 (NMM Draughts, Box 21). Since there appears to be no complete contemporary set of deck plans for a two-decker of this period, there can be no certainty that this interpretation is correct.

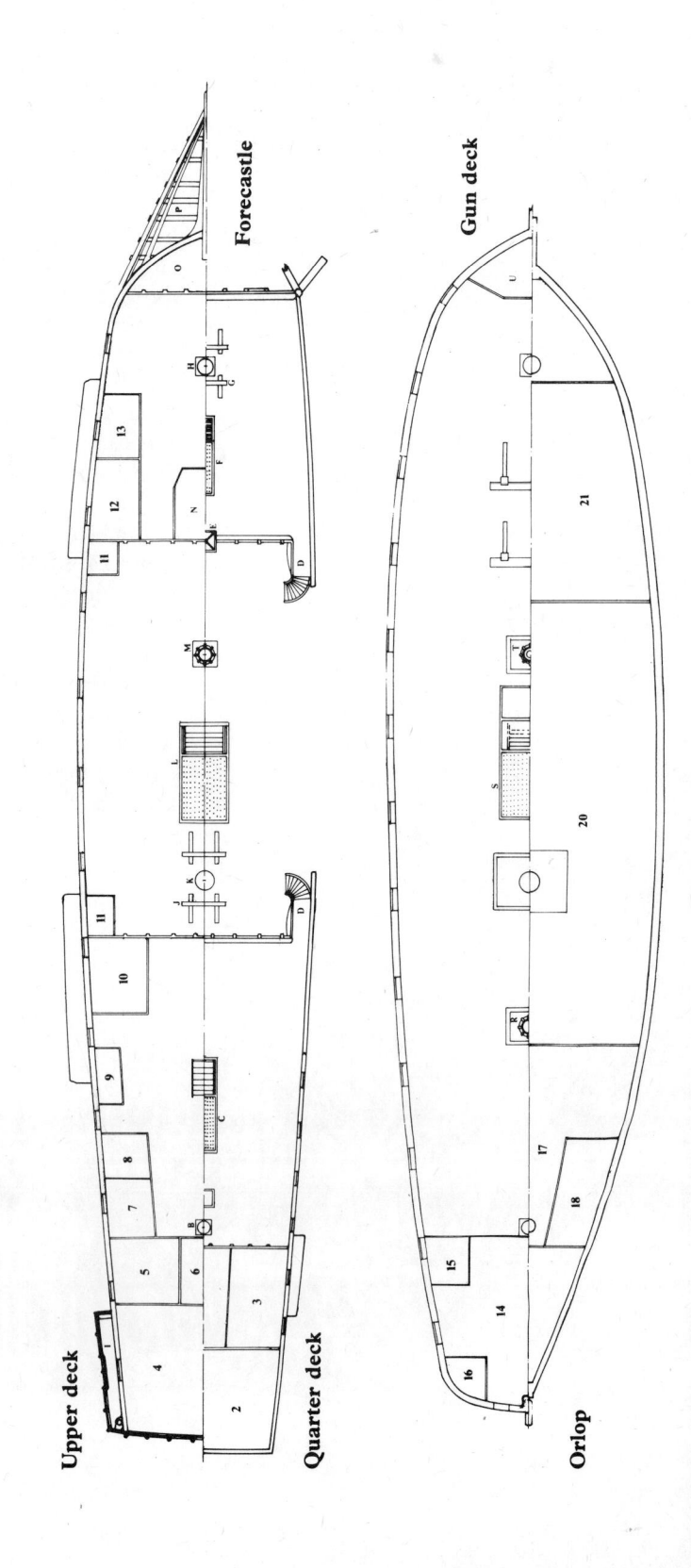

Cabins
1. Trumpeters
2. Master starboard, Lieutenant port
3. Chief Mate or 2nd Lieutenant starboard, Minister (chaplain) port
4. Great cabin, for Captain
5. Captain's bedplace
6. Captain's pantry
7. Second Mate and 'land' (marine?) officer
8. Second Mate and Pilot
9. Half cabins for servants
10. Carpenter starboard, Boatswain port
11. Coxwain and Midshipmen, under the gangways
12. Cook starboard, Boatswain's Mate port
13. Midshipman and Carpenter's Mate
14. Gunroom
15. Gunner and Surgeon
16. Standing cabins (probably for supernumerary officers)
17. Cockpit
18. Purser's cabin and Steward's storeroom and bedplace
19. Surgeon's and Captain's storerooms
20. Cable tiers
21. Boatswain's, Carpenter's and Gunner's storerooms

Deck fittings
a. Grating
b. Mizzen mast and partners
c. Gratings and ladderways
d. Gangway
e. Belfry
f. Gratings
g. Jeer bitts
h. Foremast and partners
i. Quarter gallery
j. Bitts
k. Mainmast and partners.
l. Gratings and ladderways
m. Fore jeer capstan
n. Cookroom
o. Beakhead deck
p. Beakhead
q. Tiller
r. Main capstan
s. Gratings and ladderway
t. Fore capstan
u. Manger

The upper decks of a 1st Rate, c1680.
(Pepysian Library)

3. MIDSHIP SECTION

Based on the data given in Deane's table of scantlings, with reference to Dummer's cross sections of a 1st Rate (Pepys Library Ms no 2934) The left hand side shows the floor timber and the futtocks above it, and the floor rider, and the right hand side shows the adjoining row of futtocks, and the futtock rider. (*N.B.* Twice scale)

1. Keel
2. Floor timber
3. First Futtock
4. Second Futtock
5. Third Futtock
6. Toptimber
7. Kelson
8. Orlop deck beams
9. Lower deck beams
10. Lower deck carlines
11. Lower deck ledges
12. Lodging knee
13. Upper deck beams
14. Upper deck carlines
15. Upper deck ledges
16. Lodging knee
17. Footwaling
18. Thicker strakes over rungheads
19. Footwaling to cover the join of the first and third futtocks.
20. Clamps of the orlop deck
21. Orlop deck planking
22. Orlop deck waterways
23. Spirketting
24. Gun deck hanging knee
25. Gun deck clamp
26. Orlop deck standard
27. Gun deck plank
28. Gun deck waterways
29. Spirketting
30. Upper deck hanging knee
31. Upper deck clamp
32. Upper deck beams
33. Upper deck waterway
34. Spirketting
35. Floor rider
36. Futtock rider
37. Cross pillars in hold
38. Knees
39. Upright pillar in hold
40. Stanchions
41. Underwater planking
42. Thicker planking below wales
43. Lower wales
44. Channel wales
45. Great rail
46. Gunwale
47. Limber board

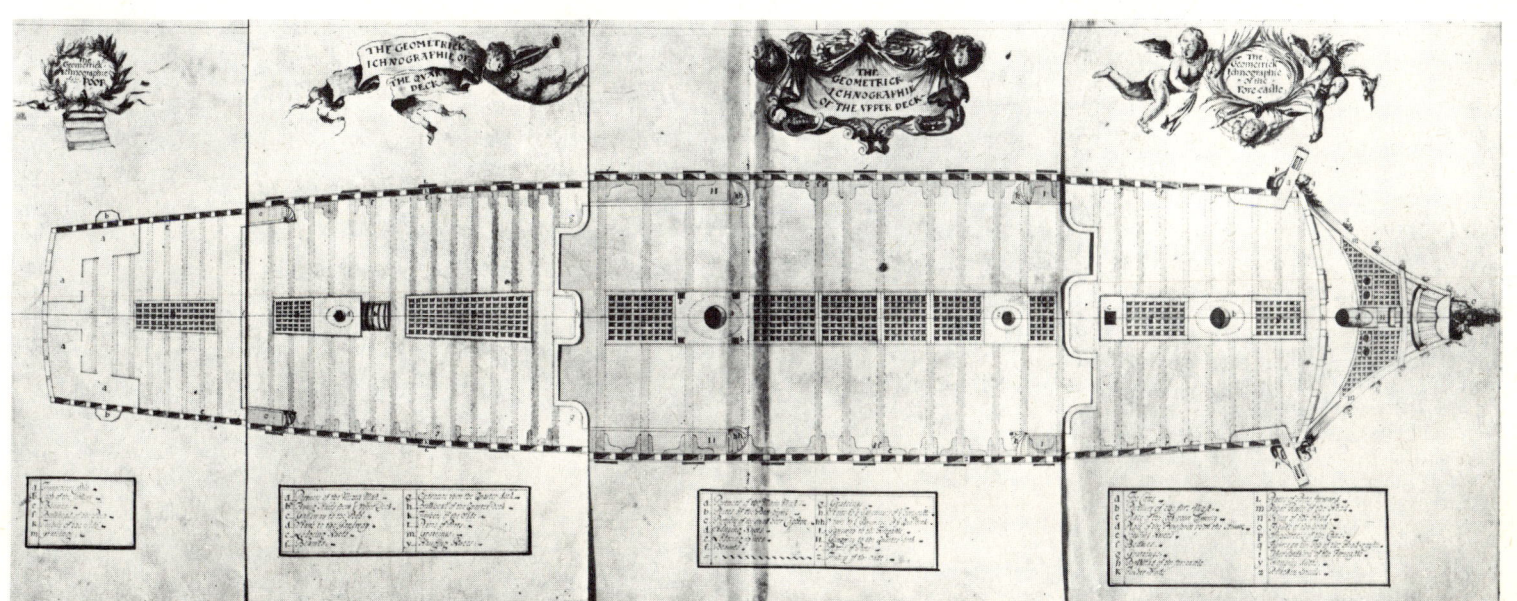

DEANE'S DOCTRINE OF NAVAL ARCHITECTURE, 1670

4. STANDING RIGGING
1. Gammoning (wolding) for bowsprit
2. Shrouds and backstay for spritsail topmast
3. Shrouds and lanyards of foremast
4. Forestay
5. Collar of forestay
6. Lanyard of forestay
7. Shrouds and lanyards of fore topmast
8. Backstays and lanyards of fore topmast
9. Fore topmast stay
10. Fore topmast lanyard
11. Futtock shrouds of fore topmast
12. Fore topgallant shrouds
13. Fore topgallant stay
14. Main shrouds and lanyards
15. Main stay, lanyard and collar
16. Shrouds and lanyards of main topmast
17. Backstays and lanyards of main topmast
18. Stay and lanyard of main topmast
19. Futtock shrouds of main topmast
20. Shrouds and lanyards of main topgallant mast
21. Main topgallant stay
22. Shrouds and lanyards of mizzen mast
23. Stay and lanyard of mizzen mast
24. Futtock shrouds of mizzen topmast
25. Shrouds and lanyards of mizzen topmast
26. Mizzen topmast stay

These rigging plans are based mostly on the information given in the *Doctrine*, especially from the rigging lists and plans, and the list of spar dimensions, but there are some notable inconsistencies in these. There is no mention of sheets for the spritsail topsail, for example. Staysails have not been shown, because Deane makes no mention of the rigging for them in his lists, though it is clear that they were in common use by 1670. Some of the gaps in the *Doctrine* have been filled by reference to other sources, notably James Lees' *The Masting and Rigging of English Ships of War, 1625-1860*.

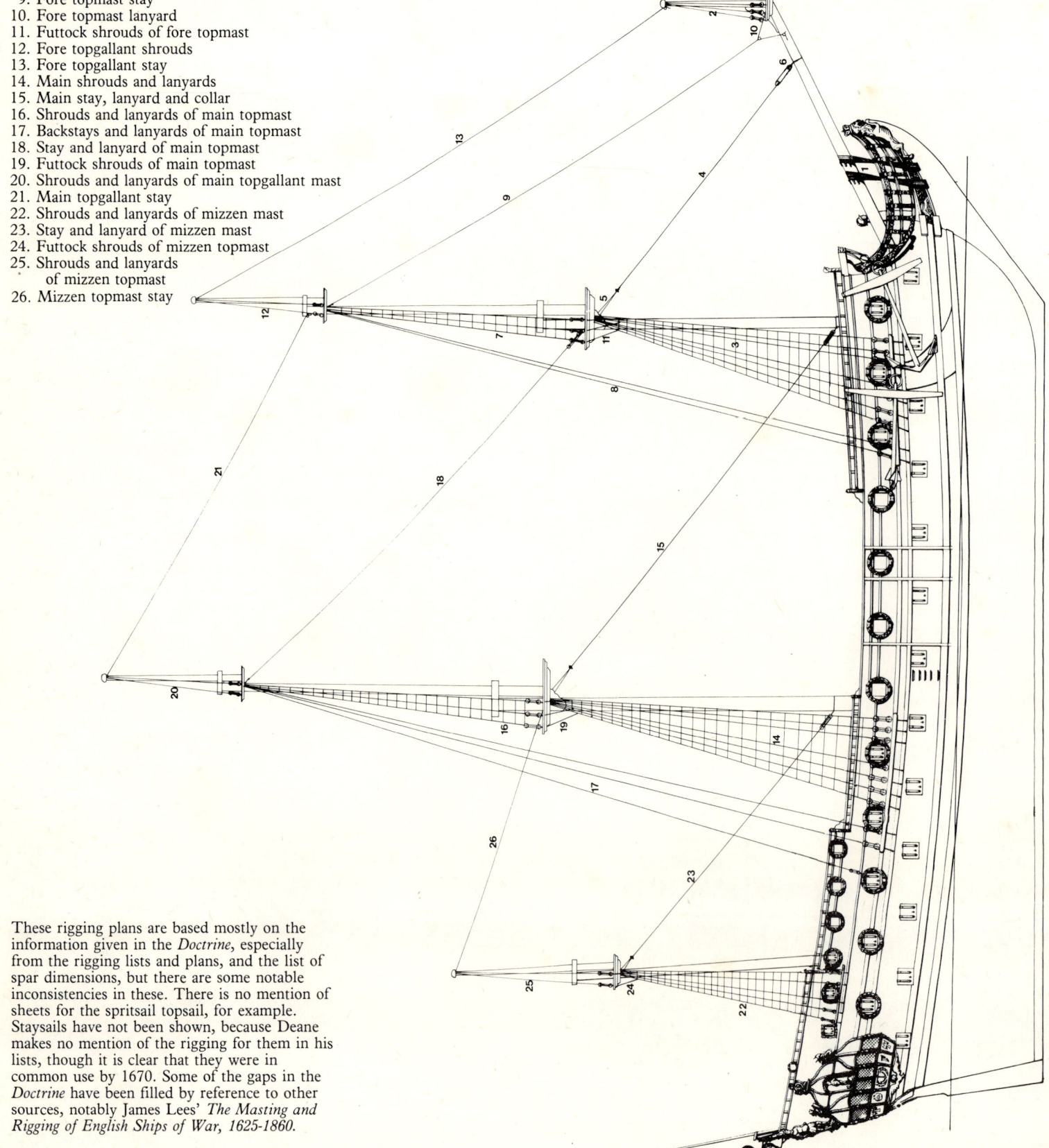

126

5. RUNNING RIGGING

Bowsprit
1. Halyard
2. Sling
3. Braces
4. Sheets
5. Horse

Spritsail topmast
6. Parrell
7. Tie
8. Braces
9. Sheets

Foremast
10. Jeers
11. Parrell
12. Braces
13. Sheets
14. Tacks
15. Bowlines and bridles

Fore topmast
16. Halyard
17. Parrell
18. Tie
19. Braces
20. Sheets
21. Bowlines and bridles

Fore topgallant mast
22. Halyard and tie
23. Parrell
24. Braces
25. Sheets
26. Bowlines and bridles
27. Pendant, guy and fall of garnet

Mainmast
28. Jeer
29. Parrell
30. Braces
31. Sheets
32. Tacks
33. Bowlines and bridles

Main topmast
34. Halyard
35. Parrell
36. Tie
37. Braces
38. Sheets
39. Bowlines and bridles

Main topgallant mast
40. Halyard and tie
41. Parrell
42. Braces
43. Sheets
44. Bowlines and bridles

Mizzen mast
45. Halyards
46. Parrells
47. Sheets
48. Tacks and bowlines
49. Brails

Crossjack yard
50. Slings
51. Braces

Mizzen mast
52. Halyard
53. Parrell
54. Braces
55. Sheets
56. Bowlines and bridles

6. MAINMAST

A. Bonnet
B. Lacing of bonnet
C. Main course
D. Clews
E. Bowline cringles
F. Topsail
G. Reef points
H. Topgallant sail

Main course
1. Clew garnets
2. Lifts
3. Leech lines
4. Buntlines

Main topsail
5. Clew lines
6. Lifts
7. Leech lines
8. Buntlines

Main topgallant sail
9. Clew lines
10. Lifts

Standing rigging
11. Main shrouds
12. Main topmast shrouds
13. Main topmast futtock shrouds
14. Catharpins
15. Main topgallant shrouds

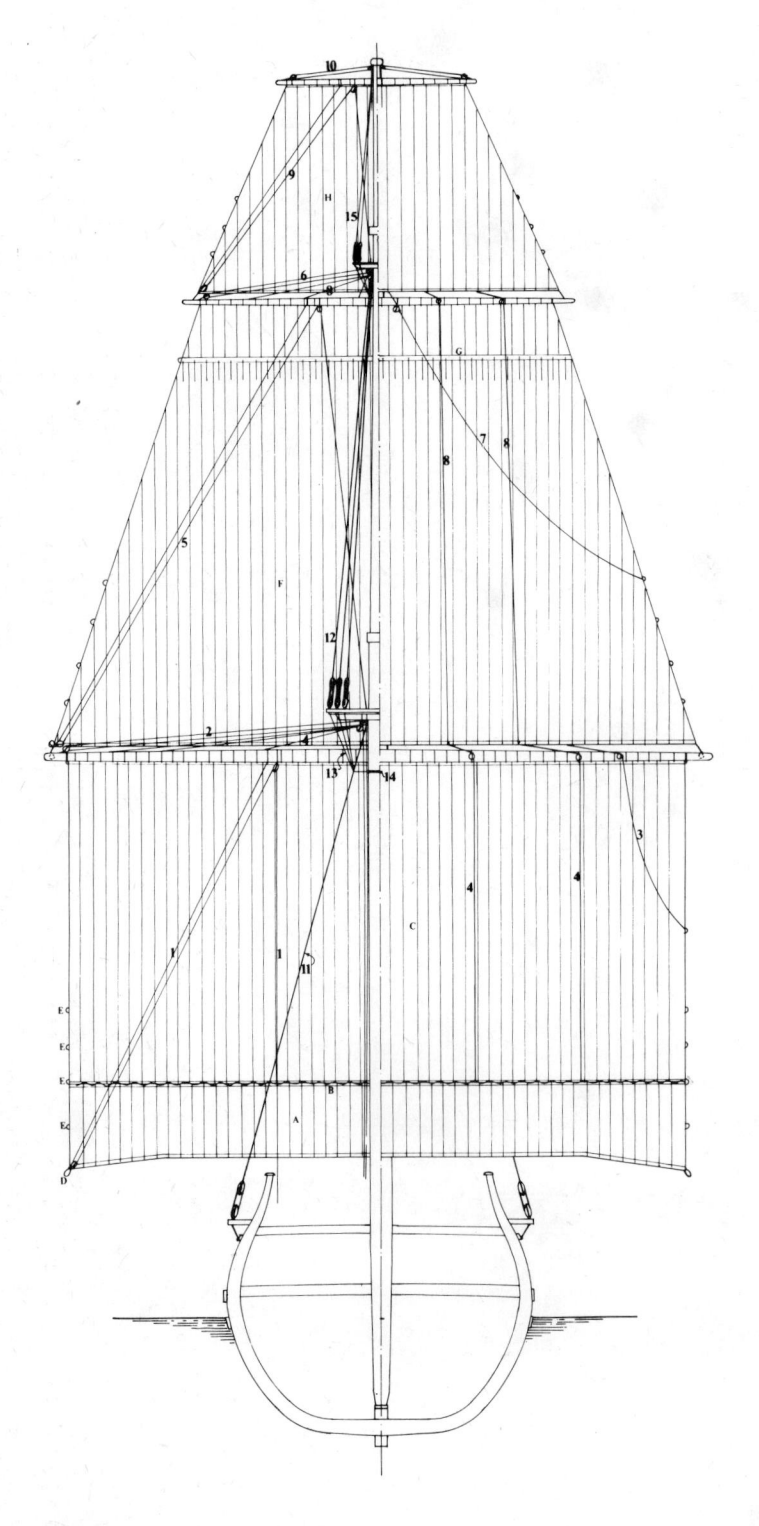

7. SPRITSAIL AND TOPMAST

A. Spritsail
B. Bowsprit
C. Spritsail yard
D. Spritsail topsail
E. Spritsail topmast
F. Spritsail topsail yard

1. Spritsail lifts
2. Spritsail clew line
3. Spritsail buntline
4. Spritsail topsail lifts
5. Spritsail topsail clew lines

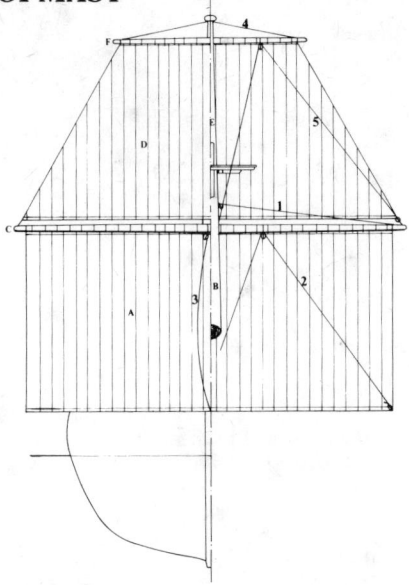

This is probably the earliest detailed drawing of sails in existence. It can be dated from slightly later than 1670, since no bonnets are carried, and these went out of use around 1680. The single row of reef points in the course, and double row on the topsail, suggests a date of between 1688 and 1710, The handwriting, and the arrangement of this particular item within a set of other papers, suggests that it dates from the beginning rather than the end of that period. *(British Library)*

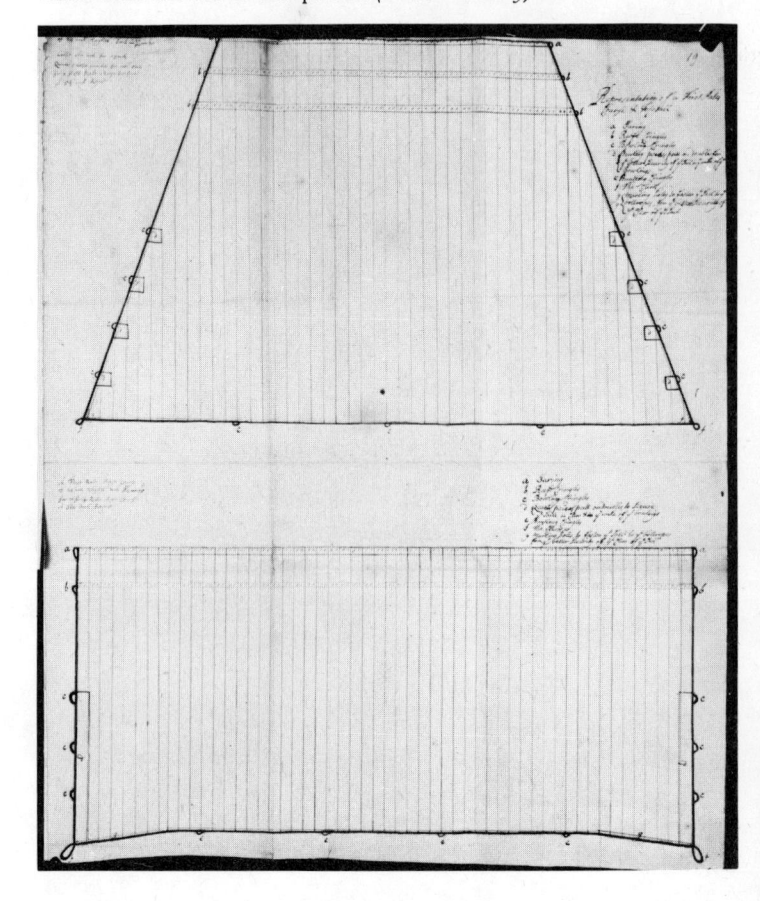